CONCEPTS IN

ENTERPRISE RESOURCE PLANNING

Joseph A. Brady
University of Delaware

Ellen F. Monk
University of Delaware

Bret J. Wagner
Western Michigan University

**COURSE
TECHNOLOGY**
™
THOMSON LEARNING

Australia • Canada • Mexico • Singapore • Spain • United Kingdom • United States

Concepts in Enterprise Resource Planning

by Joseph A. Brady, Ellen F. Monk, Bret J. Wagner

Managing Editor
Jennifer Locke

Senior Vice President, Publisher
Kristen Duerr

Production Editor
Karen Jacot

Developmental Editor
DeVona Dors

Associate Product Manager
Matthew Van Kirk

Editorial Assistant
Janet Aras

Marketing Manager
Toby Shelton

Text Designer
Books by Design

Cover Designer
Efrat Reis

Disclaimer
Course Technology reserves the right to revise this publication and make changes from time to time in its content without notice.

SAP is a registered trademark.

ISBN 0-619-01593-4

■ Dedications

To my students, with admiration

— Joe Brady

In loving memory of my parents, Christina and John Fischer

— Ellen Monk

To my parents, Charles and Audrey Wagner; my wife, Cindy; and my children, Emily, Robert, and Stuart

— Bret Wagner

CONTENTS

Preface

This is a book about Enterprise Resource Planning (ERP) systems; it's also about how a business works and how information systems fit into business operations. Why not write a book solely about ERP software? We wrote this book as we did, because teaching ERP merely as software did not work for us. If you use our book to teach ERP, we hope you will avoid the problems we encountered and therefore better educate your students.

Three years ago, the University of Delaware's College of Business and Economics became a member of SAP America's University Alliance Program. As members, we were allowed to use SAP R/3 and its training database in our classes. We eagerly looked forward to teaching SAP R/3. Finally! We had a way to *show* our students the power of integrated information systems. We could show our students data sharing in real time throughout a company's functional areas—increasing the efficiency of operations and helping managers to make better decisions. We could show how ERP can solve the problems that arise from having unintegrated information systems.

It soon became apparent that our early classes were not well received. Students viewed our SAP R/3 classes merely as tedious data-entry exercises. Where did we go wrong? After talking things over with our students, we concluded that our teaching was flawed because it was based on these faulty assumptions:

1. We assumed that all of our students understood how businesses and functional areas operate; in fact, many of our undergraduate students did not yet have a good grasp of how profit-making organizations operate.

2. We assumed that all of our students understood the problems inherent in unintegrated systems; in fact, even most of our advanced undergraduate and MBA students did not truly grasp what goes on in real companies where people in different functional areas must work together to achieve company goals.

3. We assumed that all of our students understood how an information system should help business managers make decisions; in fact, some of our students did not understand this well.

Because our approach had been to teach solutions to business problems—problems that students had yet to understand—it was no wonder our students were not connecting with us. We knew we had to take some steps back to the basics.

Our first step was to find a textbook that could help us. We found that there were many books (especially about SAP R/3) written *by* ERP practitioners *for* ERP practitioners. Unfortunately, these books were too technical or assumed the reader understood how businesses operate. We found that there were no basic "principles of ERP" texts available. So, we set out to write a book that does the following:

1. Describes basic business functional areas and explains how they are related

2. Illustrates how unintegrated information systems fail to support business functions and business processes that cut across functional area boundaries

3. Demonstrates how integrated information systems can help a company to prosper by improving business processes and by providing business managers with accurate and up-to-date data

■ The Approach of This Book

Our book describes the activities of the fictitious Fitter Snacker Company, a manufacturer of nutritious snack bars. The company's functional area information systems are somewhat primitive and unintegrated, causing many operational problems. We have intentionally made the systems' problems easy to understand, so the student can readily comprehend them. Your reaction might be to ask, "Do companies often have *that* many organizational and information system problems?" Our answer is this: A single manufacturer with all of Fitter Snacker's problems might not stay in business very long, but the problems we describe are modeled on situations we have seen in business.

This is an overview of our book's structure:

- Chapter 1 explains the purposes for, and information systems requirements of, **main business functional areas**—marketing and sales, production and materials management, accounting and finance, and human resources. This chapter also describes how a business process cuts across the activities within business functional areas and why managers need to think about making business processes work.

- Chapter 2 provides a short history of business computing and the **developments that led to today's ERP systems**. Chapter 2 concludes with an overview of ERP issues.

- Chapter 3 describes the **marketing and sales functional area**, and it highlights the problems that arise with unintegrated information systems. To make concepts easy to understand, Fitter Snacker is used as a sustained example. After explaining FS's problems with its unintegrated systems, we show how ERP can avoid these problems. SAP's R/3 screens are used to illustrate the concepts. Because using ERP can naturally lead a company into ever-broadening integration, a discussion of customer-relationship management (CRM) concludes the chapter.

- Chapter 4 describes **production and materials management** (purchasing) at Fitter Snacker. As in Chapter 3, the company's problems are explored, and then the problems are solved using ERP (again with SAP R/3 illustrations). Supply-chain management (SCM), a natural outgrowth of ERP systems, is also discussed.

- Chapter 5 describes **accounting and management-reporting issues** at Fitter Snacker. Once again, problems with the company's unintegrated information system are presented and then solved, using ERP.

- Chapter 6 reveals how ERP is evolving to respond to developments in the **World Wide Web and e-commerce**.

How Can You Use This Book?

This book is flexible and can be used to suit an instructor's needs and student population. It might be used in the following ways:

- The book, or selected chapters, could be used for a three-week ERP treatment in undergraduate Management Information Systems, Accounting Information Systems, or Operations Management courses.

- Similarly, the book or selected chapters could be used in MBA courses, such as foundation Information Systems or Operations Management courses. Although the concepts presented here are basic, the astute instructor can build on them with more sophisticated material to challenge the more-advanced MBA student. Many of the exercises in the book require research for their solution, and the MBA student could do these in some depth.

- The book could serve as an introductory text in a course devoted wholly to ERP. It would provide the student with a grounding in how ERP systems help companies to integrate different business functions. The instructor might use Chapter 6 as the starting point for teaching the higher-level strategic implications of ERP and related topics, such as supply-chain management and e-commerce. The instructor can pursue these and related topics using his or her own resources, such as case studies and current articles.

As noted, we don't feel it's a good idea to assume too much about what our students know or what classes they have taken. Except for a computer literacy course, we assume no particular educational or business background. Chapters 1 and 2 lay out most of the needed business and computing groundwork, and the rest of the chapters build on that.

Features of This Text

To bring ERP concepts to life (and down to earth!) this book uses sales, manufacturing, purchasing, and accounting examples for the Fitter Snacker Company. Thus, the student can see problems, not just at an abstract level, but within the context of a company's operations. As a result, we think this will make business problems and their ERP solutions easier to understand.

The book's exercises have the student analyze aspects of Fitter Snacker's information systems in various ways. The exercises vary in their difficulty; some can be solved in a straightforward way, and others require some research. Not all exercises need to be assigned. This gives the instructor flexibility in choosing which concepts to emphasize and how to assess students' knowledge. Some exercises explore FS's problems, and some ask the student to go beyond what is taught in the book and to research a subject. A solution might require the student to generate a spreadsheet, perform calculations, document higher-level reasoning in a memorandum, present the results of research in writing, or participate in a debate.

The book has two other features designed to bring ERP concepts to life: (1) short, detailed case studies that focus on problems faced by real-world companies, and (2) interviews with information systems managers who share their experiences with ERP.

We have illustrated ERP by showing how SAP R/3 would handle the problems discussed in the book. Screen shots of key SAP R/3 steps are shown throughout to illustrate ERP concepts. Many of the book's exercises ask the student to think about how a problem would be addressed using SAP R/3.

Supplemental Materials

This book is accompanied by the following materials:

- Instructor's manual, which includes problem solutions, sample answers, grading guidelines for exercises, sample syllabi, and lecture notes for each chapter

- Test Bank

- PowerPoint presentation slides

- Content for Blackboard, WebCT, and Course Technology's online course enhancement platform, *MyCourse.com*

■ Acknowledgments

Our thanks go out to our managing editor, Jennifer Locke, who made all the right calls as this book developed in the last year; to our tireless and long-suffering developmental editor, DeVona Dors, whose questions and comments have greatly improved our work; to SAP America's Dan Pantaleo, Amelia Maurizio, Joseph Spitz and John Kruszewski, for initial encouragement and help throughout. We also thank our reviewers Dr. Edieal Pinker, University of Rochester; Michael H. Ensby, Clarkson University; and Dr. Richard Orwig, Washington State University, for insightful comments that pointed to needed improvements. In addition, we thank our interviewees, Patricia Davis and Claudio Spiguel, for their time and frankness. And finally, we thank our students, whose honesty and desire to learn have inspired us.

CHAPTER 1

Business Functions, Processes, and Data Requirements

LEARNING OBJECTIVES

After completing this chapter, you should be able to:

- Name a business's main functional areas of operation.

- Differentiate a business process from a business function.

- Identify the kinds of data that each main functional area produces.

- Identify the kinds of data that each main functional area needs.

- Define integrated information systems and state why they are important.

Introduction

Enterprise Resource Planning (ERP) software programs are at the cutting edge of information systems technology. ERP (pronounced "E-R-P") programs help to manage company-wide business processes, using a common database and shared management reporting tools. ERP software supports the efficient operation of business processes by integrating business activities, including sales, marketing, manufacturing, accounting, and staffing. Today's business people (and tomorrow's) should understand what an ERP program can do for a company.

■ Functional Areas and Business Processes

To understand ERP, you must first understand how a business works. Let's begin by looking at a business's areas of operation, called **functional areas of operation**.

Functional Areas of Operation

Companies that make products to sell have the following main functional areas of operation. Each functional area comprises a variety of **business functions**, business activities within that functional area of operation.

1. *Marketing and Sales*: This functional area includes the business functions of marketing, sales order processing, customer relationship management, customer support, sales forecasting, and advertising.

2. *Production and Materials Management*: This functional area includes the business functions of purchasing, receiving, transportation/logistics, production scheduling, manufacturing, and plant maintenance.

3. *Accounting and Finance*: This functional area includes the business functions of financial accounting, cost allocation and control, planning and budgeting, and cash-flow management.

4. *Human Resources*: This functional area includes the business functions of recruiting and hiring, training, payroll, and benefits.

Historically, businesses have had organizational structures that separated functional areas, and business schools have been similarly organized, so each functional area is taught as a separate course. Thus, one might conclude that what happens in one functional area is not closely related to what happens in others. Similarly, one might conclude that one area's **information system (IS)** has no impact on those of other functional areas. As you will learn in this chapter, however, functional areas are interdependent, each requiring data from the others.

2

Business Processes

Recently, managers have begun to think in terms of business processes rather than business functions. A **business process** is a collection of activities that takes one or more kinds of input and creates an output that is of value to the customer. This is illustrated by the diagram shown in Figure 1.1.

FIGURE 1.1 A business process

Thinking in terms of business processes helps managers to look at their organization from the customer's perspective. For example, suppose that a customer wants to purchase a new computer. She wants information about the company's products so she can select a computer and options. She wants to place her order quickly and easily, and perhaps arrange for financing through the company. She expects quick delivery of a correctly configured, working computer, and she wants 24-hour customer support for any problems. The customer is not concerned about how the computer was marketed, or how its components were purchased, or how it was built, or the route the delivery truck took to get to her house. The customer wants the satisfaction of having a working computer at a reasonable price.

So, what's the difference between a business function and a business process? Suppose the customer's computer is damaged during shipment. It's a *business function* of Customer Service to accept the damaged item. The actual repair and redelivery of the item, however, is a *business process* that involves several functional areas and functions within those areas. Thus, the customer is looking *across* the company's functional areas in her process of buying and obtaining a computer. Business managers are now trying to view their business operations from the perspective of a satisfied customer.

For the computer company to provide customer satisfaction, it must make sure that its functional areas of operation are integrated. For example, computer technology changes rapidly, and the hardware the computer company offers changes frequently. Therefore, people performing the Sales function must have up-to-date information about computer configurations so they can provide the customer with accurate information; otherwise, the customer might order a computer system that the company's manufacturing plant no longer produces. Similarly, people performing the Manufacturing function need to get the details of the customer's computer configuration quickly and accurately from those performing the Sales function, so the right computer can be manufactured and shipped on time to the customer. If the customer is financing the computer through the computer company, then those performing the Sales Order Processing function must gather information about the customer and process it quickly, so financing can be approved in time to support shipping the computer. In addition, if the customer has a technical problem with the computer after delivery, then people performing the Customer Support function also need information about the configuration the customer ordered.

Sharing data effectively and efficiently between and within functional areas leads to more efficient business processes. Information systems can be designed so that accurate and timely data are shared between functional areas. These systems are called **integrated information systems**. This textbook's goal is to help you understand the benefits of integrated information systems and the problems that can occur when information systems are not integrated. Figure 1.2 illustrates the process view of business operations.

FIGURE 1.2 A process view of business

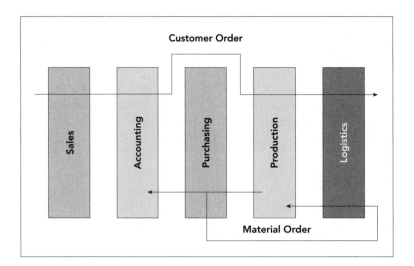

Businesses take inputs in the form of material, people, and equipment and transform these inputs into goods and services for the customer. Information is required to manage these inputs and the business processes so that the customer receives the goods and services requested. For example, a customer's order is processed by Sales, and the product's manufacture is scheduled by Production. Logistics schedules and carries out the product's delivery. If raw materials are needed to make the product, Production would prompt Purchasing to arrange for their purchase and delivery. In that case, Logistics would receive the material, verify its condition to Accounting so that the vendor could be paid, and deliver the goods to Production. Throughout, Accounting would keep appropriate transaction records.

As an introduction, we'll look at how business processes involve more than one functional area, using a very small business as an example—a lemonade stand that you might have operated as a child. We'll examine the business processes of the lemonade stand and see why coordination of functional areas is required to achieve efficient and effective business processes. You'll see the role that information plays in this coordination and how integration of the information system improves efficiency and effectiveness.

■ Functional Areas and Business Processes of a Very Small Business

FIGURE 1.3 Basic business processes are important even in a very small business.

Even though one person can run a lemonade stand, the operation of the business requires a number of processes. Coordinating the activities within different functional areas requires accurate and timely information.

Marketing and Sales

The functions within Marketing and Sales include developing products, determining pricing, promoting products to customers, and taking customers' orders. Marketing and Sales also helps to create a sales forecast that will be used to ensure the successful operation of the stand.

For the most part, this is a cash business, but you still need to keep track of your customers, so that you can send flyers to repeat customers or send occasional "thank you" notes. Thus, your records must show not only the amount of sales, but also identify repeat customers.

Product development can be done informally in such a simple business—you gather information about who buys which kind of lemonade and note what customers say about each product. You also analyze historical sales records to spot trends that are not obvious. Deciding whether to sell a product also depends on how much it costs to produce the product. For example, some customers might be asking for a new sugar-free lemonade at a certain price. Data from Production and Materials Management could be analyzed to see whether the lemonade could be profitably produced and sold.

Even though you run a cash business, very good repeat customers are allowed to "charge" purchases—*up to a point*. Thus, the records must show how much each customer owes and his available credit. It's very important that the data are available and accurate at the time of a customer's credit request. Thus, Accounting records must be accessed as a part of the selling process, so the Accounting function has a critical role to play in the sales process.

Production and Materials Management

Functions within Production and Materials Management include making the lemonade and buying raw materials. Production is planned so that, as much as possible, lemonade is available when needed, avoiding excess production of lemonade that must be "liquidated." This requires sales forecasts from the Marketing and Sales functional area, the accuracy of which will be improved if forecasts are based on historical sales figures. Thus, forecasts from Marketing and Sales play an important role in the Production Planning Process.

In addition, production plans are used to develop orders for raw materials (bottled spring water, fresh lemons, and raw sugar). The raw material orders are based on the production plan, so sufficient raw materials are available to support the production plan, but excessive inventory that might spoil is not carried.

Production Planning must choose a recipe for each lemonade product sold and communicate this information to Purchasing. The standard recipe is a key input for deciding how much to order of each raw material. This also implies a need to keep good manufacturing records, so managers within Production and Materials Management (working with those in Accounting and Finance) can later compare how much it costs to make a glass of lemonade, versus how much the recipe should have cost.

Accounting and Finance

Functions within Accounting and Finance include recording raw data about transactions, including sales, raw material purchases, payroll, and receipt of cash from customers. Those data are then summarized in meaningful ways to determine the profitability of the lemonade stand and to support decision making.

Note that data from Accounting and Finance are used by Marketing and Sales as well as by Production and Materials Management. The sales records are an important component of the sales forecast, which is used in making staffing decisions and in production planning. The accounts receivable records, which you use to determine whether to grant credit to a particular customer, are also used to monitor the overall credit-granting policy of the stand. You want to be sure that you have enough cash on hand to purchase raw materials, as well as to finance new equipment purchases, such as a lemon juicer.

Human Resources

Even in a simple business, employees must be found to support the Sales and Production functions. This implies a need to recruit, train, evaluate, and compensate employees.

At the lemonade stand, the number and timing of employees to hire depends on the level of lemonade sales, so sales forecasts will be used to determine Human Resource (HR) needs. A part-time helper might be needed at forecasted peak hours or days. How much should a part-time helper be paid? That depends on prevailing job market conditions, and it's HR's job to monitor those conditions.

Would increased sales justify hiring a part-time worker at the prevailing wage? Or, should you think about acquiring more automated ways of making lemonade, so that one person working alone could run the stand? These questions must be studied with input from Production and Materials Management and from Accounting and Finance.

The lemonade stand, while a simple business, has many of the processes needed in larger organizations. These processes involve activities in more than one functional area; in fact, it is impossible to discuss the processes in one functional area without discussing the linkages to other functional areas. These linkages invariably require sharing data.

■ Functional Area Information Systems

The lemonade stand provides simple examples of business processes and the functional activities required to support them. Next we will expand on this example to describe potential inputs and outputs for each functional area. Note the kinds of data needed and how people use the data. Also note the information systems relationships between each functional area.

Marketing and Sales

Customers communicate their orders to Marketing and Sales by telephone, e-mail, fax, the Web and such. In the case of Web-based systems, data should be stored automatically in the information system; otherwise, data must be stored manually, using keyboard input or point-of-sale systems, bar code readers, and so on. Sales orders must be passed to Production and Materials Management for planning purposes and to Accounting and Finance for billing. In addition, sales order data can be used to analyze sales trends for business decision making. For example, Marketing and Sales management might use a report showing the trend of a product's sales to evaluate marketing efforts and to determine strategies for the sales force.

Marketing and Sales also has a role in determining product prices, which requires an understanding of the market competition and the costs of manufacturing the product. This might be done based on a product's unit cost, plus some percentage markup. For example, if a product costs $5 per unit to make, and management wants a 40% markup, the selling price must be $7 per unit. Where does the per-unit cost come from? Determining the cost of manufacturing a product requires information from Accounting and Finance which, in turn, relies on Production and Materials Management data.

FIGURE 1.4 The Marketing and Sales functional area exchanges data with customers and with the Human Resources, Accounting and Finance, and Production and Materials Management functional areas.

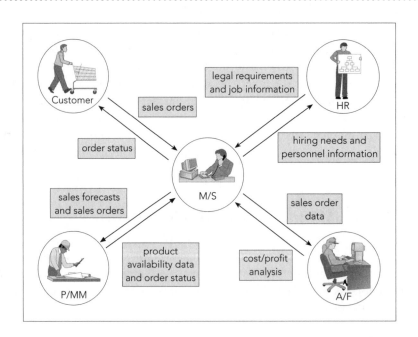

Production and Materials Management

Manufacturing firms develop production plans of varying length and detail.

- Long-range plans, which may span more than one year, are used to plan capacity expansions (e.g., new buildings and equipment) and personnel needs (e.g., hiring and training of employees).

- Medium-range plans, which may span a number of months, are used for planning material needs for long-range production and for ordering purchased materials (order quantities and timing). In addition, these plans encompass manufacturing capacity (hiring temporary employees or paying for overtime) and production quantities for groups of products. These plans are based on sales forecasts and on-hand inventory.

- Short-range plans are usually determined from the medium-range plans and deal with individual products rather than product groups. These plans can also account for such changes as unexpected orders, employee absences, and material delivery problems.

Production plans are based on information about product sales (actual and projected) that comes from Marketing and Sales. The Purchasing function bases its orders of raw materials on production plans, expected shipments, delivery lead-times, and existing inventory levels. With accurate data about required production levels, raw materials can be ordered as needed, and inventory levels can be kept low, saving money. On the other hand, if data are inaccurate or not current, Manufacturing may actually run out of raw material (called a "stockout"). Such a shortage will shut down production and might cause the company to miss delivery dates. To avoid stockouts, management might carry extra raw materials, known as safety stock, which could result in an over-investment in inventory. Accurate Production information can frequently reduce the need for extra inventory and manufacturing capacity.

Production and Materials Management records can provide the data needed by Accounting and Finance to determine how much was spent (raw materials, labor, supplies, and overhead) to make finished-goods inventory.

Production and Materials Management data can support the Sales function by providing information about what has been produced and shipped. For example, some computer manufacturers, such as Gateway, have automated systems that call customers to notify them that their computer order has been shipped. Shipping companies, like UPS and Federal Express, provide shipment-tracking information via the Internet. By entering a tracking number, the customer can see each step of the shipping process by noting where the package's bar code was scanned. Thus, accurate and timely production information can support the sales process and increase customer satisfaction.

For long-range planning, management might want to see monthly reports showing sales and production figures. The data for such reports must come from the production and inventory data.

FIGURE 1.5 The Production and Materials Management functional area exchanges data with suppliers and with the Human Resources, Marketing and Sales, and Accounting and Finance functional areas.

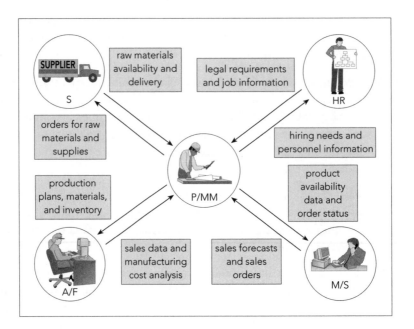

Accounting and Finance

Accounting and Finance personnel record the company's transactions in the books of account. For example, they record accounts receivable when sales are made and cash receipts when customers send in payments. In addition, they record accounts payable when raw materials are purchased and cash outflows when they pay for materials. Finally, Accounting and Finance personnel summarize the transaction data to prepare reports about the company's financial position and profitability.

People in other functional areas provide data to Accounting and Finance: Marketing and Sales provides sales data, Production and Materials Management provides production and inventory data, and Human Resources provides payroll and benefit expense data. The accuracy and timeliness of Accounting and Finance data depend on the accuracy and timeliness of Marketing and Sales *and* Production and Materials Management data.

Marketing and Sales personnel require data from Accounting and Finance to evaluate customer credit. If a customer's order will cause him to exceed his credit limit, Marketing and Sales should see that the customer's accounts receivable balance is too high and hold new orders until the customer's balance is lowered. If Accounting is slow to record sales or customers' payments, then accounts receivable balances are inaccurate. As a result, credit might be approved for customers who have already exceeded their credit limit and may never pay off their account. Similarly, it would also be possible to deny credit to customers who actually owe less than their credit limit, potentially damaging the relationship with the customer.

FIGURE 1.6 The Accounting and Finance functional area exchanges data with customers and with the Human Resources, Marketing and Sales, and Production and Materials Management functional areas.

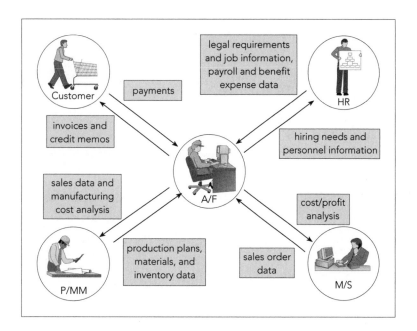

Human Resources

To properly recruit new employees, people in Human Resources (HR) need accurate forecasts of personnel needs. HR also needs to know what skills are needed to perform a particular job and how much the company can afford to pay employees. The data come from all functional units.

State and federal laws require companies to observe many governmental regulations in recruiting, training, compensating, promoting, and terminating employees—and these regulations must be observed company-wide. Usually, it is also HR's responsibility to assure that employees receive training in a timely manner. HR also must assure that employees get certified (and recertified) in key skills, such as materials handling and equipment operation. In addition, employees must be paid, and they must be considered for raises and bonuses. For these and other reasons, corporate HR needs timely and accurate data from other areas.

HR must create accurate and timely data for management use. For example, HR should maintain a database of skills required to do particular jobs and the prevailing pay rate. The data can be used to evaluate employees' performance and compensation. This analysis can help to prevent losing valued employees because of low pay.

FIGURE 1.7 The Human Resources functional area exchanges data with the Accounting and Finance, Marketing and Sales, and Production and Materials Management functional areas.

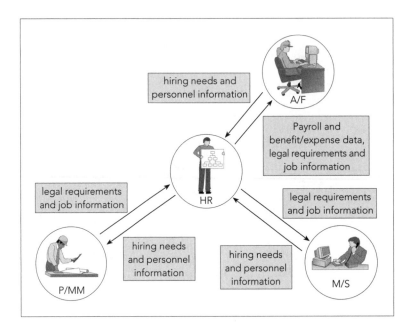

Fortunately, ERP software allows all functional areas to use a single common database so that accurate, "real time" information is available. In the next chapter, we will trace the evolution of data management systems that has led to ERP.

■EXERCISES

After reading this chapter, you should be able to answer the following questions:

1. Distinguish between a business function and a business process. Provide an example to illustrate your points. Why do managers today try to think in terms of business processes rather than business functions?

2. Describe one situation in which information needs to be shared between Marketing and Accounting. What is the business purpose? What kinds of data are shared?

3. Describe one situation in which information needs to be shared between Production and Marketing. What is the business purpose? What kinds of data are shared?

4. Describe one situation in which information needs to be shared between Production and Human Resources. What is the business purpose? What kinds of data are shared?

SUMMARY

1. Companies that make products to sell have business processes that involve these basic functional areas: Marketing and Sales, Production and Materials Management, Accounting and Finance, and Human Resources. They perform these functions:
   - Marketing and Sales sets product prices, promotes products, takes customer orders, and creates sales forecasts.
   - Production and Materials Management develops manufacturing plans, orders raw materials from suppliers, receives the raw material into the facility, manufactures products, and ships products to customers.
   - Accounting and Finance records sales transactions, records customers' payments, records suppliers' invoices and payments to suppliers, and summarizes operational data in managerial reports.
   - Human Resources recruits, trains, compensates, and oversees the evaluation of employees.

2. A functional area is served by an information system. Information systems capture, process, and store data to provide information needed for decision making.

3. Employees working in one functional area need data from those in other functional areas. Ideally, functional area information systems should be integrated, so shared data are accurate and timely.

4. Today business managers try to think in terms of business processes that integrate functional areas. This way of thinking promotes efficiency and competitiveness. An important aspect of this is the need to share information between functions and functional areas. ERP software encourages this by using a single common database.

CHAPTER **2**

The Development of Enterprise Resource Planning Systems

LEARNING OBJECTIVES

After completing this chapter, you should be able to:

- Identify the factors that led to the development of Enterprise Resource Planning (ERP) systems.

- Describe the distinguishing characteristics of ERP software.

- Discuss the pros and cons of implementing an ERP system.

- Summarize ongoing developments in ERP.

Introduction

In today's competitive business environment, companies try to provide customers with goods and services faster and cheaper than their competition. How do they do that? Often, the key is to have efficient, integrated information systems. Increasing the efficiency of information systems results in more efficient management of business processes. When companies have efficient business processes, they can be more competitive in the marketplace.

Until recently, most companies had unintegrated information systems that only supported the activities of individual business functional areas. Thus, a company would have a Marketing information system, a Production information system, and so on, each with its own hardware, software, and methods of processing data and information.

Such unintegrated systems might work well within individual functional areas, but to achieve a company's goals, data must be shared throughout functional areas. When a company's information systems are not integrated, costly inefficiencies can result. For example, suppose two functional areas each have their own unintegrated information system. To share data, a clerk in one functional area needs to print out data from another functional area and then type the information into her area's information system. Not only does this data input take twice the time, it also doubles the chance for data entry errors. Alternatively, the process might be automated by having one information system write data to a file to be read by another information system. This would reduce the probability of errors, but it could only be done periodically (usually overnight or on a weekend) to minimize the disruption to normal business transactions. Because of the time lag in updating the system, the transferred data would rarely be up to date.

By contrast, ERP programs can integrate a company's operations by acting as a company-wide computing environment that includes a shared database. An ERP system's impact on costs and operational efficiency can be startling. For example, IBM Storage Systems division reports that it can now reprice its product inventory in five minutes rather than five days. Shipping a replacement part now takes three days instead of 22 days. Doing a customer credit check takes three seconds instead of 20 minutes.[1] These dramatic improvements in efficiency can lead to lower costs and more satisfied customers.

This chapter will help you to understand how and why ERP systems came into being and what the future might hold for business information systems. The chapter follows this sequence:

- A review of the evolution of information systems reveals why ERP systems were not possible until recently.

- ERP's distinguishing characteristics and parts of an ERP system—its "architecture"—are described.

- A few ERP software vendors dominate the market, and they are noted. The current industry leader, German software maker SAP AG and its revolutionary software product R/3 are discussed.

- Factors influencing a company's decision whether to purchase an ERP are reviewed.

- ERP systems are the subject of numerous questions. These frequently asked questions are dealt with as a group.

- ERP software currently represents the state of the art in business software, but information technology is always changing. The chapter concludes with a discussion of the future of ERP software and its impact on the Internet and e-commerce.

■ The Evolution of Information Systems

It seems obvious today that a business should have integrated software to manage all functional areas. An integrated ERP system, however, is an incredibly complex hardware and software system that was not feasible until the 1990s. Current ERP systems evolved as a result of (1) the development of hardware and software technology needed to support the system and (2) the development of a vision of integrated information systems.

Computer Hardware and Software Development

The first computers were large and expensive and had less computing capability than a modern handheld calculator. The first large-scale automatic digital computer, Howard H. Aiken's Harvard Mach I, was built in 1944. The computer was 50 feet long, 8 feet tall, and weighed 5 tons. It consisted of 750,000 components that included switches, relays, shafts, and clutches.[2]

The First Business Computers
The first practical business computers were the mainframe computers of the 1960s. These computers were housed in their own climate-controlled rooms for operational and security purposes. Programs were entered using punched cards (each line of a computer program was written on a separate card) or on magnetic tape. Running a program meant giving it to a technician who physically loaded the cards or tape into the machine, where the media might wait for hours before actually being processed by the computer. Long, complicated programs were frequently run at night when the demand on the computer was low.

There was no way these early computer systems could provide integrated, real-time data for business decision making. These mainframe computers were only suitable for performing sequential, repetitious data-processing tasks, such as accounting, inventory management, and payroll. As a result, companies purchased mainframe computers to do isolated, repetitious business tasks that previously required a large numbers of clerks. Thus, because of technological limitations, early information systems could not be integrated, even if the value of integration had been understood.

Computers Get Smaller and Faster

Computers got smaller, cheaper, and faster during the 1960s and 1970s. Storage and retrieval technology improved to allow direct, or random, access to data on magnetic disks, rather than the slower sequential access that characterized punched cards and tape. All programs that accessed data immediately became faster as a result of this change.

Relational databases were introduced in the 1970s and benefited from the change in storage technology. Relational databases were better at answering complex queries (e.g., "How many units of our products did each salesperson sell last month in our three sales regions?") than were hierarchical databases, but relational databases required faster computers and disk access.

Computer hardware became even smaller in the 1980s. The microprocessor ("computer on a chip") became the electronic hobbyist's toy in the 1970s, but matured into a serious business tool when IBM introduced the first personal computer (PC) in 1981. The increasing capability and decreasing cost of the personal computer helped make it a standard business tool.

The dramatic increase in computer performance is described by **Moore's Law**, named after Gordon Moore. In 1965 Moore noticed that a new generation of computer chips was being released every 18 to 24 months, and the number of transistors on a computer chip (and thus its performance) doubled with each new generation. As Figure 2.1 shows, his observation has continued to hold true. The exponential growth of computer power, which is now assumed, continues to radically transform information systems.

FIGURE 2.1 The actual increase in transistors on a chip approximates Moore's Law. Data Sources: *HowStuffWorks.com* and *Webopedia.com*

Year Introduced	Intel Chip	Transistors
1972	4004	2,300
1974	8080	6,000
1979	8088	30,000
1982	80286	134,000
1985	80386	275,000
1989	80486	1,200,000
1993	Pentium 1	3,100,000
1997	Pentium 2	7,500,000
1999	Pentium 3	10,000,000

Software Proliferates

The release of personal productivity software, primarily word-processing software and spreadsheet software, helped to make PCs popular. Word-processing software allowed business people to type their own documents, rather than relying on clerks. Changes to documents could be made quickly and easily. In addition, standard documents could be created, then slightly altered to fit new circumstances. In 1979 the first spreadsheet package for the PC, Visicalc, was introduced. Spreadsheet software provided users with the power to develop complex business analyses without programming.

For the first time, these two applications gave business people control over their own computing. As more business people joined the PC revolution, companies realized they needed to connect individual users' PCs and integrate their data.

Early Attempts to Share Data

As PCs gained popularity in business, managers became aware that important business information was being stored on individual PCs, but there was no easy way to share the information electronically. It was easy to configure a PC to function as a "dumb terminal" for a mainframe computer, but what was needed was a way to share expensive peripheral equipment (like printers and hard disks, which in the early 1980s were fairly expensive) and, more importantly, data.

Developments in telecommunications allowed data and peripherals to be shared on "local" networks. This meant that data could be downloaded from a central computer (now often called a "**server**") to the local PC (now often called a "**client**" of the server) to do the actual work with the data. Client-server architecture began to take the place of mainframe-based architecture. Servers became more powerful and less expensive (following Moore's Law) and provided **scalability**. Scalability means that when a client-server network's capacity is exceeded, its capacity can be increased by adding a new, additional server—thus extending the life of the hardware investment. By contrast, when a mainframe system's capacity is exceeded, the entire system must be replaced.

By the end of the 1980s, much of the hardware needed to support the development of ERP systems was in place: fast computers, networked access, and advanced database technology. In addition, the database and software tools required to manage the development of complex ERP software existed. What was needed was the vision to see the benefits of integrated information systems as well as the willingness to commit the resources to develop ERP software.

The Manufacturing Roots of ERP

The vision of having an integrated information system began on the factory floor. Manufacturing software developed during the 1960s and 1970s, evolving from simple inventory tracking systems to **Materials Requirements Planning (MRP)** software. MRP software allowed a plant manager to plan production and raw materials requirements by working backward from the sales forecast. Thus, the manager (1) looked at Marketing and Sales' forecast of demand, (2) looked at the Production schedule needed to meet that demand, (3) calculated the raw materials needed to meet production, and (4) then projected raw materials purchase orders to suppliers. For a company with many products, raw materials, and shared production resources, this kind of computation was impossible without a computer.

The basic functions of MRP could be handled by mainframe computers using sequential file processing. **Electronic data interchange (EDI)** allowed for companies to handle the purchasing process electronically, avoiding the cost and delays resulting from paper purchase order and invoice systems. What we now call **supply-chain management (SCM)** began with the sharing of long-range production schedules between manufacturers and their suppliers.

The basic concept of the ERP system was inherent in the development of **MRP II-Manufacturing Resource Planning**. The term "MRP II" was coined by Oliver Wight, a key figure in the development of MRP systems during the early 1980s. As Wight explained:

> Instead of having one set of numbers for the operating system in manufacturing and another set kept by the financial people, once the manufacturing people have numbers that are valid, the financial people can use these to get their numbers. Of course, whenever there are two systems, the numbers are bound to be different. With MRP II, everybody can be working to the same set of numbers. But that's only the technical difference. The big difference comes in the way management uses these tools. . . . Manufacturing Resource Planning becomes a company game plan for manufacturing, marketing, engineering, and finance.[3]

Thus, it is possible to view ERP systems as an extension of MRP II.

Management's Impetus to Adopt ERP

Companies' downsizing and reorganization efforts in the late 1980s and early 1990s, which were brought on by hard economic times, were also a stimulus to ERP development. Companies needed to find some way to avoid the following kind of situation, which they had been living with for a long time.

Imagine you're the CEO of a clothing manufacturing company. Your company is profitable and is keeping pace with the competition. Your IS is unintegrated and inefficient, as are those of your competitors. You've learned to live with this kind of inefficiency: Your Sales and Marketing department creates a time-consuming paper trail for negotiating and making a sale. To schedule factory production, however, your Manufacturing manager needs accurate, timely information about actual and projected sales orders from the Sales and Marketing manager. Without such information, the Manufacturing manager must guess which products to produce—and how many of them to produce. To keep goods moving through the production line, she often does guess. Sometimes she overestimates demand for some garments, and sometimes she underestimates demand.

Overproduction of a certain garment might mean your company is stuck with garments for which there is no market, OR a diminishing market due to style changes or seasonal demand.

When you store the garments, waiting for a buyer, you incur warehouse expense. On the other hand, underproduction of a certain garment might result in garments not being ready for delivery when a salesperson promised, resulting in unhappy customers and cancelled orders. If you try to "catch up" on orders, you'll have to pay factory workers overtime pay, or resort to the extra expense of rapid-delivery shipments.

The management of large companies decided they could no longer afford this kind of in efficiency—the result of a functional model of business organization. This model had deep roots in U.S. business, starting with the General Motors organizational model developed by Alfred P. Sloan in the 1930s. That model was very useful for decades. In the rapidly changing markets of the 1990s, however, the model led to top-heavy and overstaffed organizations incapable of reacting quickly to change. The time was right to view a business as a set of cross-functional processes. Michael Hammer's 1993 landmark book, *Reengineering the Corporation: A Manifesto for Business Revolution,* stimulated managers to see the importance of managing business processes. Books like Hammer's and the difficult economic times led to a climate in which ERP software was seen as a solution to business problems.

There are a number of ERP software vendors; however, we will describe the development of ERP software using the SAP company. SAP was first to develop ERP software and is the current market leader in ERP software sales.

■ ERP Software Emerges: SAP and R/3

In 1972 five former IBM systems analysts formed *Systemanalyse und Programmentwicklung* (Systems Analysis and Program Development—pronounced "S-A-P") in Mannheim, Germany. In their work for IBM, they noticed that many clients wanted computer programs to manage the same business functions—financial accounting, production planning, sales order entry, and so forth. The analysts had these goals for their new software company:

- They wanted to develop and market standard software that would integrate business processes.

- They wanted data to be processed interactively and to be available in real time.

- They wanted users to work with data on a computer screen, and not with voluminous printed output.

These goals were lofty and forward-looking for 1972, and it took almost 20 years to achieve them.

SAP Begins Developing Software Modules

SAP founders took the first step toward achieving their goals by acquiring the rights to a batch mode (non-real time) financial accounting package. Using this package as a starting point, they developed a standard, real-time financial accounting package, selling the first copy to a German subsidiary of ICI, a large chemical company headquartered in the U.K. As sales (and the reputation) of the Financial Accounting package grew, SAP developed its second standard product, a Materials Management program with modules for Purchasing, Inventory Management, and Invoice Verification. Data from the Materials Management program could flow easily into the Financial Accounting program.

By 1977 SAP had become a closely held corporation (GmbH). It moved its headquarters to the small town of Walldorf, Germany, and changed its name to *Systeme, Anwendungen, Produkte in der Datenverarbeitung*—in English, "Systems, Applications, and Products in Data Processing." SAP added another central module, Asset Accounting, to the product line and developed a French version of the Financial Accounting module, starting the company on the road to being an international software company.

Because of the continued development of mainframe computer technology, in 1978 SAP began developing a more integrated version of its software products, called the **R/2 system**. The company continued to add additional capabilities to its software, such as order tracking. By 1980, 50 of the 100 largest industrial companies in Germany were SAP customers.

Sales grew rapidly in the 1980s, and SAP developed additional modules for Cost Accounting, Production Planning and Control, Personnel Management, and Plant Maintenance. In addition, expansion into international markets was increased. This was no small task, because the software had to be able to accommodate different currencies, accounting practices, and tax laws.

By 1988 SAP GmbH had established subsidiaries in numerous foreign countries, established SAP Consulting GmbH as a joint venture with consulting company Arthur Andersen, and sold its 1,000th system to Dow Chemical. SAP also became SAP AG, a publicly traded company.

SAP's R/3 Release

In 1988 SAP realized the potential of client-server hardware architecture and began development of its **R/3 system**, which was released in 1992. Each subsequent release of R/3 software contained new features and capabilities. SAP's client-server architecture allowed it to run on a variety of computer platforms, including Unix and Windows NT. The R/3 system was also designed using an **open-architecture** approach. This means that third-party software companies were encouraged to develop add-on software products that could be integrated with SAP R/3. The architecture also makes it easy for companies to integrate their hardware products—like bar code scanners, PDAs (personal digital assistants), cell phones, and global information systems (GIS)—as well as their software with the R/3 system.

The company continued to grow rapidly. The approaching millennium spurred sales because many companies decided to solve their Y2K data management problems by converting to the R/3 system. This led to a dramatic increase in business in the late 1990s, but that was followed by a business slump in 1999 as many companies, which had not yet decided to move to a Y2K-compliant ERP system, waited until after the new millennium to upgrade their information systems.

By 2000 SAP AG had 22,000 employees in 50 countries—and 10 million users at 30,000 installations around the world. Its main competitors in the ERP market are PeopleSoft, J.D. Edwards, Oracle, and Baan.

SAP's R/3 system (and those by other vendors) allows all business areas to access the same database, as shown in Figure 2.2, eliminating redundant data and communications lags. For example, when a salesperson enters a sales order into the database, those data are immediately available to Production so manufacturing the order can begin and raw materials can be restocked. The data are also then available to Accounting so an invoice can be prepared.

FIGURE 2.2 Data flow within an integrated information system

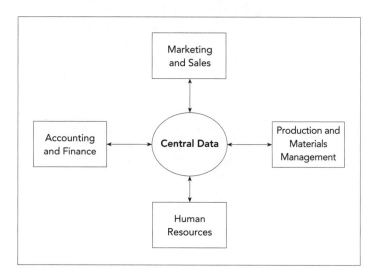

Figure 2.3 is SAP's way of showing this integration. The diagram's central diamond represents the company-wide computing environment, including the centralized database, the heart of the R/3 system. The labeled boxes surrounding the central diamond represent the basic modules available in SAP's R/3 integrated package. The empty cells represent possible third-party software add-ons.

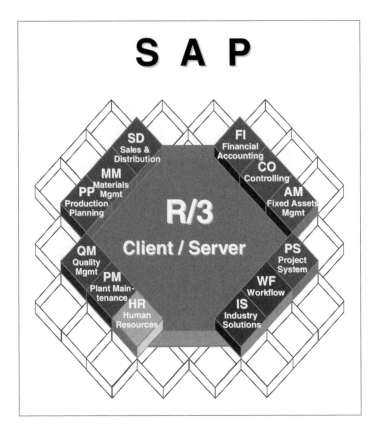

Notice how modules are arranged in Figure 2.3. Adjacent modules have closely related functions. The modules on the left side of the diagram are for Marketing and Sales, Production and Materials Management, and Human Resources:

- The **Sales and Distribution (SD)** module records sales orders and scheduled deliveries. Information about the customer (pricing, how and where to ship products, how the customer is to be billed, etc.) is maintained and accessed from this module.

- The **Materials Management (MM)** module manages the acquisition of raw materials from suppliers (purchasing) and the subsequent handling of raw materials inventory, from storage to work-in-progress goods to finished-goods inventory.

- The **Production Planning (PP)** module maintains production information. Here production is planned and scheduled, and actual production activities are recorded.

- The **Quality Management (QM)** module helps to plan and record quality-control activities, such as product inspections and material certifications.

- The **Plant Maintenance (PM)** module allows planning for preventative maintenance of plant machinery and managing maintenance resources, so equipment breakdowns are minimized.

- The **Human Resources (HR)** module facilitates employee recruiting, hiring, and training. This module also includes payroll and benefits.

The modules shown on the upper-right side of the diagram are for financial management:

- The **Financial Accounting (FI)** module records transactions in the general ledger accounts. It also generates financial statements for external reporting purposes.

- The **Controlling (CO)** module is used for internal management purposes. Here, the company's manufacturing costs are assigned to products and to cost centers, facilitating cost analysis.

- The **Asset Management (AM)** module helps the company to manage fixed-asset purchases (plant and machinery) and the related depreciation.

- The **Project System (PS)** module allows the planning for and control over new R&D, construction, and marketing projects. This module allows for costs to be collected against a project, and it is frequently used to manage the implementation of the SAP R/3 system.

The two modules on the lower-right side of the diagram are not modules in the same sense as the others:

- The **Workflow (WF)** module can be used to automate any of the activities in R/3. It can perform task-flow analysis and prompt employees (via e-mail) if they need to take action.

- The **Industry Solutions (IS)** module contains R/3 configuration settings that SAP has found are appropriate for particular industries. These settings simplify R/3 implementation and let the buyer take advantage of SAP's industry experience.

To summarize: ERP integrates business functional areas. Before ERP, each functional area operated independently, using its own information systems and ways of recording transactions. ERP software also makes management reporting and decision making faster and more uniform throughout an organization. In addition, ERP promotes thinking about corporate goals, as opposed to thinking only about the goals of a single department or functional area.

SAP R/3 Software Implementation

A truly integrated information system entails integrating all functional areas, but for various reasons, not all companies install all R/3 modules. For example, a company without factories wouldn't choose the Manufacturing modules. Another company might consider its HR department's operations so separate from its other operations, it might decide not to integrate its HR functional area. Another company might feel that its internally developed production and logistics software gives it a competitive advantage, and so it may choose to implement Finance and Accounting plus Human Resource modules—but integrate its internally developed ("legacy") production and logistics system into the R/3 system.

Generally, the level of data integration is highest when one vendor supplies all of a company's modules. When modules from different vendors are chosen, additional work is needed to bridge their differences. Thus, a company must be sure the decision to use multiple vendors or to maintain a legacy system is based on sound business analysis, not on a resistance to change.

After a company chooses its major modules, it chooses from an assortment (about 8,000) of "configuration options." These options are shown in configuration tables, which are like a menu from which the company customizes (to some degree) the modules it has chosen. For example, in the Accounting module, a business might choose to track product costs and revenue by factory, by product, by geographical region, and so on.

Unique Features of SAP's R/3

Not only was SAP's R/3 software the first software that could deliver enterprise systems integration, it was and is unique in other ways:

1. SAP's original R/3 system targeted very large companies. Formerly, it was assumed these giants could never be integrated because of the sheer amount of computing power required to integrate them. Increased computing speeds, however, meant that large companies in a variety of industries, including manufacturing, gas and oil, airlines, and consulting, could have integrated information systems.

2. Acquiring an R/3 software system is very expensive. In addition, the software is so powerful, many companies find they must buy new hardware to accommodate the software. For a Fortune 500 company, software, hardware, and consulting costs can easily exceed $100 million. Full implementation can take years.

3. The modular design of R/3 is based on business processes, such as sales order handling, materials requirement handling, and employee recruiting. So, when data is entered, a ripple effect is created, automatically updating data in all related files in the central database. No human input is required to make the changes.

4. In earlier eras, information systems people felt that software should be designed to reflect a business's practices. As SAP accumulated experience developing information systems, however, the company began to develop models of how certain industries' business processes should be managed in a way compatible

with the R/3 system. Thus, R/3's design incorporates "best practices," which means that R/3 designers choose the best, most efficient ways in which business processes *should* be handled. If a buyer's business practices don't follow R/3's design, then *the business must redesign its practices so it can use the software.* Basically, R/3 is more than a software package—it's a way of doing business.

By 1998, most of the big Fortune 500 companies had already installed ERP systems, so ERP vendors refocused their marketing efforts on midsize companies (those with fewer than 1,000 employees).[4] Midsize companies represented a ripe, profitable market. For example, midsize European companies have a total yearly budget of $50 billion for IT expenditures. The American market is even larger.[5] To appeal to this new market, SAP developed industry-specific preconfigured versions of R/3 that cost less and could be installed more quickly. Application hosting, in which a third-party company provides the hardware and support rather than an internal IS department, is also making ERP systems like SAP more appealing to midsize companies.

"Best of Breed" Approaches

It's worth noting that all vendors' ERP software is not the same. For example, PeopleSoft reputedly offers the strongest Human Resources package, while SAP is renowned for its Production Planning and Materials Management. Some companies have opted to integrate modules from different ERP vendors, a design approach called "Best of Breed."

The Best of Breed approach may rapidly become obsolete, however. With a global development force of almost 6,000 people, the SAP R/3 system is constantly evolving. In response to PeopleSoft's perceived advantage in HR, SAP put a significant amount of development work into its R/3 4.5 release's HR module.

Similarly, SAP has noted complaints about the difficulty of implementing the R/3 system. (There has been a cancelled implementation by Dell Computers, a lengthy implementation at Owens-Corning, and a lawsuit by the now-bankrupt FoxMeyer drug company.[6]) In response, SAP has developed the **Accelerated SAP (ASAP)** implementation methodology to ease the implementation process.

In addition, SAP continues to extend the capabilities of R/3. For example, a **Customer Relationship Management (CRM)** module can now be included to help manage a mobile sales force. As this module continues to be improved, there will be less incentive for companies to buy CRM from other vendors to integrate into R/3. Thus, it's important to evaluate not just what an ERP vendor's software does today, but what its future development plans are as well.

Choosing Consultants and Vendors

Because ERP packages are so large and complex, one person can't fully understand a single ERP system—or compare various systems. So, before choosing a software vendor, most companies study their needs and then hire a team of consultants to help choose the right software vendor(s) and the best approach to implementing ERP. Working as a team, the consultants apply their expertise to selecting an ERP vendor (or vendors) that best meets their customer's needs.

After selecting a vendor, the consultants recommend which modules are best suited to the company's operations, and which configurations within those modules are most appropriate. This preplanning should involve not only the consultants and a company's IT department, but management as well.

The Significance and Benefits of ERP Software and Systems

The significance of ERP lies in its many benefits. As was previously mentioned, integrated information systems can lead to more efficient business processes that cost less than those in unintegrated systems. ERP systems have these benefits as well:

1. ERP allows easier global integration: Barriers of currency exchange rates, language, and culture can be bridged automatically, so data can be integrated.

2. ERP not only integrates people and data, but it also eliminates updating and repairing many separate computer systems. For example, Boeing had 450 data systems that fed data into its production process. The company now has a single way to record production data.[7]

3. ERP allows management to manage operations, not just monitor them. For example, without ERP, getting an answer to "How are we doing?" requires getting data from each business unit and then putting the data together for a comprehensive, integrated picture. The ERP system already has all the data, allowing the manager to focus on "What are we going to do better?" This enhances management of the company as a whole, and makes the organization more responsive when change is required.

■ Questions About ERP

How Much Does an ERP Package Cost?

Cost includes several factors:

- ERP software. The cost for the entire implementation is often predicated on the size of the software package, which corresponds to the size of the company it must serve.

- New hardware that's capable of running complex ERP software

- Consultants' and analysts' fees

- Time for implementation (disruption of business)

- Training (costs both time and money)

The Large Company

A large company might spend from $50 million to $500 million for an ERP system. For example, consider Allied Signal, Inc., a $14.5 billion-per-year manufacturer. The company is implementing SAP R/3 in its turbocharging systems division, which encompasses 11 countries, 18 sites, and 9 languages.[8] Because its business empire is vast, the company's ERP installation will be complex and the database huge. Such an installation might cost as much as $30 million in software license fees, $200 million in consulting fees, additional millions to purchase new hardware, and even more millions to train managers and employees.[9] In addition, full implementation of the new system might take four to six years.

The Midsize Company

A midsize company (one with fewer than 1,000 employees) will spend far less than a large company. Typically, such a company might spend $10 million to $20 million in total implementation costs and have its ERP system up and running in about two years.

Should Every Business Buy an ERP Package?

ERP packages imply, by their design, a way of doing business, and they require users to follow that way of doing business. Some of a business's operations, and some segments of its operations, might not be a good match for an ERP package's requirements. Therefore, it's imperative for a business to analyze its business strategy, organization, culture, and operation *before* choosing an ERP approach.

An article in the *Harvard Business Review* provides examples that show the value of planning: "Applied Materials gave up on its system when it found itself overwhelmed by the organization changes involved. Dow Chemical spent seven years and close to half a billion dollars implementing a mainframe-based enterprise system; now it has decided to start over again on a client-server version."[10]

It may also be the case that a company is not ready for ERP. In many cases, ERP implementation difficulties result when management does not fully understand its current business processes and cannot make implementation decisions in a timely manner.

While many people claim that ERP systems, especially the SAP R/3 system, are rigid, R/3 does offer numerous configuration options. In addition, programmers can write specific routines for special applications in SAP's internal programming language, called **Advanced Business Application Programming (ABAP)**.

So, Is ERP Software Inflexible?

Once an ERP system is in place, trying to reconfigure a system while retaining data integrity is expensive and time-consuming. That's why thorough pre-implementation planning is so important. It's much easier to customize an ERP program when the system is being configured and no data has been stored.

In addition, ERP does not always readily lend itself to companies that experience rapid growth and change. For example, in 1997 Dell Computer cancelled most of a two-year-old

R/3 project when tests showed the software couldn't handle anticipated sales volume.[11] At that time, Dell's operation was growing by $1 billion every six to eight weeks. The company's rapid growth would have meant constant and costly updating. In addition, the company didn't feel that any single vendor offered all the options they wanted. So, Dell implemented flexible "middleware" software so they could buy software from a variety of vendors—and make changes to their data that would keep up with the company's rapid expansion.[12] Expensive? Yes. The right solution? For them, yes.

What Return Can Be Expected from an ERP Investment?

The benefits provided by an ERP system can be difficult to calculate because (1) sometimes ERP increases revenue and decreases expenses in intangible ways that are difficult to measure, and (2) some changes take place over such a long period of time, that they are difficult to track. Here are some factors affecting the return on an ERP investment:

- Because ERP eliminates redundant effort and duplicated data, there can be a savings in operations expense. One study indicated that 33% of companies saw a cost savings in sales order management, and 34% of companies said their ERP system significantly reduced their personnel needs.[13]

- Because an ERP system can help move goods and services through the pipeline faster, more sales can be generated every month.

- In some instances, a company that doesn't implement an ERP system might be wedged out of business by competitors who have an ERP system—how do you calculate the monetary advantage of remaining in business?

- A smoothly running ERP system can save a company's personnel, suppliers, distributors, and customers much frustration—a benefit that is real but difficult to quantify.

- Because both cost savings and increased revenues occur over many years, it's difficult to put an exact dollar amount accrued from the original ERP investment.

- Because ERP implementations take time, there may be other business factors affecting the company's costs and profitability, making it hard to isolate the impact of the ERP system alone.

How Long Does It Take for Most Companies to See a Return on an ERP Investment?

The **Return on Investment (ROI)** for an investment project is calculated by dividing the value of the project's benefits by the value of the project's cost. An ERP system's ROI can be difficult to calculate because of so many intangible costs and benefits, as previously mentioned. Some companies don't even try to make the calculation, on the grounds that the package is as necessary as having electricity (which is not justified as an investment project). Companies that do make the ROI calculation have seen radically different results.

Some companies claim that an acceptable ROI began to accrue almost immediately. For example, Pitney Bowes, a $4 billion-per-year manufacturer of fax machines and copiers, said that ERP cut its costs of overall operations by 28%. In addition, sales reps can give customers accurate price quotes in 15 minutes rather than several hours, which increased sales nearly 4%. (A 4% increase doesn't sound like much, but 4% of $4 billion is $160 million!) The new system has also increased sales representatives' data entry accuracy by 41%, which translates to happy customers who are getting the right products faster.[14]

Some companies wait longer to measure their ROI. Toro, a wholesale lawnmower manufacturer, spent $25 million and four years to implement an ERP system. At first, ROI was difficult to quantify. Then, the emergence of an expanded customer base of national retailers, like Sears and Home Depot, made it easier to quantify benefits. For example, a yearly savings of $10 million was due to inventory reduction—the result of better production, warehousing, and distribution methods.[15]

On the other hand, a few companies claim that high ERP costs adversely affected profits. As mentioned previously, FoxMeyer claims that its SAP implementation caused its bankruptcy.

Why Do Some Companies Have More Success with ERP Than Do Others?

There have been reports indicating that only a low percentage of companies experience a smooth rollout of their new ERP systems *and* immediately begin receiving the benefits they anticipate. Such reports need to be put into perspective. *All* kinds of software implementations can suffer from delays, cost overruns, and performance problems. This has been one of the Information Systems field's major problems since the early days of business computing. Nevertheless, it is worth thinking about why ERP installation problems can happen.

Usually, a bumpy rollout and low ROI are due to *people* problems, not computer malfunctions:

1. Some executives blindly hope that new software will cure fundamental business problems that are not curable by any software.

2. Some executives and IT managers don't take enough time for a proper analysis during the planning and implementation phase.

3. Some executives and IT managers skimp on employee education and training.

Many ERP implementation experts stress the importance of proper education and training for both employees and managers. A company's managers and employees *will resist* changing the way they've done their jobs. For example, managers at one company resisted losing their autonomy and began squabbling among themselves. In addition, the managers were slow to provide information to the ERP implementation team. To keep the

ERP implementation from being derailed, top management was forced to take more direct control of the implementation. Many analysts have noted that active top management support is crucial for successful implementation.

Some companies willingly part with funds for software and new hardware, but don't properly budget for employee training. ERP software is complex and can be intimidating at first. This fact alone argues for adequate training. Typically, ERP training costs $10,000 to $20,000 per employee.[16] Some analysts recommend allocating 11% of the project's budget for training.[17] The cost includes training employees how to use the software to do their job, employees' nonproductive downtime during training, and—very important—educating employees about how the data they control affects the entire business operation.

The cost of poor training can be very high because with an ERP system, a data entry error can't be easily corrected with a reversing keystroke or a phone call: The error creates a ripple effect throughout the business's entire operation. For example, Owens-Corning discovered that its people didn't understand the new ERP system when customers complained. When clerks entered customers' orders incorrectly, manufacturing made and shipped those orders, resulting in customers getting the wrong product or no product. When accountants entered the wrong data, it created massive problems throughout the company's entire production line. Top management said, "Stop!" and ruled that employees and managers couldn't get on the new computer system until they passed a proficiency test.[18]

Owens-Corning planners expected to devote about 6% of their total project budget to training employees how to use the new computers and software. In fact, that number will be closer to 13%. Training was so time-consuming, one plant's manufacturing unit had to shut down because the people responsible for ordering raw materials were in class.

The Owens-Corning story does have a happy ending, however. After spending four years and $100 billion on its ERP implementation, Owens-Corning is seeing a return on its investment. In 1998 ERP software helped Owens-Corning to save $50 million in logistics, materials management, and sourcing.[19]

■ The Continuing Evolution of ERP

Understanding the social and business implications of new technologies is not easy. Howard H. Aiken, the inspiration behind the first large-scale digital computer, the Harvard Mach I, predicted in 1947 that only six electronic digital computers would be needed to satisfy the computing needs of the entire United States! Hewlett-Packard passed up the opportunity to market the computer created by Steve Wozniak that became the Apple I. Bill Gates did not appreciate the importance of the Internet until 1995, by which time Netscape controlled the bulk of the Internet browser market.[20] Thus, even people who are most knowledgeable about a new technology do not always fully understand its capabilities or how it will change business and society.

ERP systems have only been in common use since the mid-1990s. As this young technology continues to mature, ERP vendors are working hard to solve adaptability problems that plague customers. Fundamental developments are discussed next.

Additional Capabilities Within ERP

Some ERP developers are continuing to develop Customer Resource Management (CRM) packages that increase the efficiency of the sales force. For example, if sales people can target the most profitable customers, actual and potential, they'll have a competitive advantage. Other software focuses on detecting changes in customer satisfaction and responding quickly to remedy any problems.

ERP developers are also trying to make their existing systems smarter by extending ERP's capabilities into more areas of decision support, management reporting, and **data mining**. Data mining is the statistical and logical analysis of large sets of transaction data, looking for patterns that can aid decision making. For example, discovering patterns in customer behavior can lead to better marketing efforts.

In addition, ERP vendors continue to improve software and Internet connections that integrate not only a business's internal operations, but also integrates the business with its dealers, vendors, and customers. You'll read more about this in Chapters 4 and 6.

The Internet

The Internet's rapid development since the late 1990s has been something of a threat to ERP software developers. ERP software lets users access the company's software and central database via internal pathways. Now there is a need for users to access that central database directly from the Internet. This has forced ERP companies to rethink how users get to and use their ERP software. ERP developers have been marrying Web-based systems with their ERP product. This effort is typified by SAP's effort, which they call *mySAP.com*.

The following are some of *mySAP.com's* many facets:

- *mySAP.com* lets a company link a Web-based sales-order system with its ERP system. Orders come in over the Web to the company's Web site and are passed on to the ERP system.

- Employees can use *mySAP.com* to personalize their interface with the company's systems. This includes allowing access to the ERP system from remote locations.

- SAP has been trying to use the Internet to form industry-wide **business-to-business (B-to-B)** trading environments. Communication between companies is via the Internet, using *mySAP.com*. Information processing is done with traditional ERP products.

The Internet has become an important way to sell goods and services and will probably become more important in the future. There will be a continuing need to take orders electronically and to pass them seamlessly to the company's database. At that point, the ERP system can manage the transaction as if it had come in via a traditional method, such as the telephone.

In e-commerce companies, much attention is paid to the Web interface. A company that intends to sell its products on the Web, however, must still manage its business processes.

While some have speculated that e-commerce will make ERP systems obsolete, this is not likely to be the case. Rather, e-commerce should be viewed as another activity that ERP systems can help manage. Chapter 6 will elaborate on the importance of the Internet.

Another Look at Consultants

Some ERP implementations are challenging in the extreme. In the September 1, 2000, issue of *CIO* magazine, Bill Jeffery and Jim Morrison, consultants for A.T. Kearney, report one such implementation.[21]

At the end of 1995, Delphi Automotive Systems Europe, then a European subsidiary of General Motors, began planning an ERP implementation. From the start, it was a challenge. The company had 3,500 employees at 58 sites in eight European countries—and dozens of incompatible manufacturing and distribution systems. Operations within each country had their own business practices and were subject to different laws and regulations. The company's goal was to spend $50 million to link its diverse operations with a single ERP system.

At the end of 1997, Delphi Automotive called in consulting company A.T. Kearney for help. Delphi wanted an ERP implementation that was completed on time, within budget, *and* showed a real ROI. A.T. Kearney accepted the job. EDS Corp. was brought in to manage the technical aspects of the SAP R/3 implementation. Both consulting companies agreed to have a stake in the success of the project: Any cost or time overruns were the responsibility of the consultants, not their client.

The situation at Dephi Europe became more difficult. In June 1999, the company split from its parent company, General Motors, which left Delphi with no IT staff. Delphi's own in-house IT expertise was described as "minimal." In addition, the threat of Y2K loomed.

How does a consultant steer a client through an implementation with so many obstacles? In the article, A.T. Kearney claims that a highly structured implementation is the key to success:

- Identify what you want to improve (better service, etc.), set goals for improvement, and calculate the benefit you expect to receive.

- Define standard procedures (accounting methods, etc.) that will be implemented company-wide. Then decide how and when to implement the procedures.

- Plan for rollout far before it happens. Then, before rollout, perform a readiness assessment to see whether transition to the ERP system is possible.

- Set milestones for each phase of the operation, track the progress of each plan, and monitor implementation schedules and costs. Absolute vigilance is required to keep the plan on time and on task.

- Train everyone involved in the project, accommodating cross-cultural differences as needed. Make sure everyone understands their job and the goals of the project.

- Don't forget your customers. A.T. Kearney advises, "During any major transformation of a company's core business processes, all changes must be absolutely transparent to customers. In Delphi's case … the slightest hiccup in manufacturing plans could have an enormous financial impact on a customer's business."[22]

So, was the implementation a success? Yes. Delphi has an integrated information system that allows the fast and accurate flow of information within the company and with its customers and suppliers. Delphi CIO Juan Sanchez says, "The team made a number of very wise decisions early in the process. We are a completely different company now."[23]

Questions

1. Given what you know about this project, what benefits do you think accrued as a result of having the ERP system?

2. A.T. Kearney and EDS were willing to accept the risk of cost overruns. They apparently felt that they could manage the project and make a profit on it, and they did. Given what you know about the project, why do you think they were so optimistic?

▓ EXERCISES

After reading this chapter, you should be able to answer the questions and do the activities that follow.

1. What major factors in the history of business and computing led to the development of ERP?

2. What are the benefits of purchasing an ERP system? What are the disadvantages of implementing an ERP system?

3. Imagine that you have a job interview with the owner of a manufacturing company that employs about 300 people. The owner of the company has heard about ERP, but she hasn't done any research on it. She asks you, "Do you think we need an ERP system?"

What would you say to her? What questions might you ask to help her think about ERP in her company? List your response and questions.

4. Much has been written about ERP, both in the popular press and on the Internet. Using library resources or the Internet, report on (1) one company's positive experience with implementing ERP and (2) one company's disappointing experience.

5. Companies whose common stock is traded on U.S. stock exchanges are required to publish an annual report to stockholders. Obtain the annual reports of SAP, Oracle, and PeopleSoft for the

past three years. Write a report comparing how the companies say their business has changed in that time. Include charts showing changes in important operational and financial results.

■ SUMMARY

1. Several factors led to the development of ERP:
   - The speed and power of computing hardware increased exponentially, while cost and size decreased.
   - The early client-server architecture provided the conceptual framework for multiple users sharing common data.
   - Increasingly sophisticated software facilitated integration, especially in two areas: Accounting/Finance and Manufacturing Resource Planning.
   - The growth of business size, complexity, and competition made business managers yearn for more efficient and competitive information systems.

2. SAP AG produced a complex, modular ERP program called R/3. The software could integrate a company's entire business by using a common database that linked all operations, allowing real-time data sharing and streamlined operations.

3. Purchasing ERP software is expensive, time consuming to implement, and requires significant employee training, but the payoffs can be spectacular. For some companies, however, the ROI may not be immediate or even calculable.

4. In the future, experts anticipate that ERP's focus will be on managing customer relationships, improving planning and decision making, and linking operations to the Internet.

■ End Notes

1. Thomas H. Davenport, "Putting the Enterprise into the Enterprise System," *Harvard Business Review* (July-August 1998): 121–31.

2. Clive Maxfield and Alvin Brown, *Bebop BYTES Back (An Unconventional Guide to Computers)* (LLH Technology Publishing, 1997).

3. Oliver Wight, *The Executive's Guide to Successful MRP II* (Oliver Wight Limited Publications, 1982).

4. Kuldeep Kumar and Jos Van Hillegersberg, "ERP Experiences and Evolution," *Communications of the ACM* 43, No. 4 (April 2000): 23–26.

5. Yvonne Van Everdingen, Jos Van Hillegarsberg, and Eric Warts, "ERP Adoption by European Midsize Companies," *Communications of the ACM* 43, No. 4 (April 2000): 27–31.

6. Davenport, "Putting the Enterprise into the Enterprise System," 121–31.

7. Polly Schneider, "Human Touch Sorely Needed in ERP," *CIO* (online), March 2, 1999.

8. Schneider, "Human Touch Sorely Needed in ERP."

9. David Kirkpatrick, "The E-Ware War," *Fortune* (December 7, 1998): 102–12.

10. Davenport, "Putting the Enterprise into the Enterprise System," 121–31.

11. Joseph B. White, Don Clark, and Silvia Ascarelli, "Program of Pain," *Wall Street Journal*, March 14, 1997.

12. Derek Slater, "How to Choose the Right ERP Software Package," *CIO* (online), February 1999.

13. Tom Stein, "Making ERP Add Up," *Information Week.com*, May 24, 1999.

14. Stein, "Making ERP Add Up."

15. Stein, "Making ERP Add Up."

16. Richard Stooker, "SAP: A Hot Skill—The Current #1 Resource Planning Software Package," Inforing Press (online), 2000.

17. Malcolm Wheatley, "ERP Training Stinks," *CIO* (online), June 1, 2000.

18. Stein, "Making ERP Add Up."

19. Stein, "Making ERP Add Up."

20. Gates, however, did dramatically reshape Microsoft around an Internet strategy by the late 1990s. Their Internet browser is now as commonly used as any other.

21. Bill Jeffery and Jim Morrison, "ERP, One Letter at a Time," *CIO* (September 1, 2000): Bill Jeffery and Jim Morrison are vice president and principal, respectively, with global management consulting company A.T. Kearney.

22. Bill Jefferey and Jim Morrison, "ERP, One Letter at a Time."

23. Bill Jefferey and Jim Morrison, "ERP, One Letter at a Time."

CHAPTER 3

Marketing Information Systems and the Sales Order Process

LEARNING OBJECTIVES

After completing this chapter, you should be able to:

- Describe the unintegrated sales processes of Fitter Snacker, a fictitious company whose operations are used as an example.

- Explain why unintegrated Sales and Marketing information systems lead to company-wide inefficiency, higher costs, lost profits, and customer dissatisfaction.

- Discuss sales and distribution in SAP's R/3 system, and explain how integrated data sharing increases company-wide efficiency.

- Describe how SAP R/3 processes a standard sales order.

- Describe the benefits of Customer Relationship Management software, a useful extension of ERP software.

Introduction

In this chapter, you will begin reading about the operations of Fitter Snacker (FS), a fictitious company that makes healthy snack bars. FS does not have an integrated information system, and throughout this book FS will be used to illustrate information systems concepts in general and ERP concepts in particular.

As is the case in many other companies, Marketing and Sales is at the center of FS's activities. Why? Because Marketing and Sales is responsible for selling the company's product. As a result, marketing personnel often guide the company's key strategies and tactics. Marketing personnel in most companies make the following kinds of decisions:

- What products should we produce?

- How much of each product should we produce?

- How are our products best promoted and advertised?

- How should our products be distributed for maximum customer satisfaction?

- What price should we charge for our products?

On a day-to-day basis, Marketing and Sales is involved in generating key transaction data, including data for recording sales, creating customers' bills, and allocating credit to customers. FS's Marketing and Sales information systems are not well integrated with the company's other information systems. This results in inefficient use of transaction data, as you will see in this chapter. You'll also see how FS's Marketing and Sales information systems could be improved with ERP. Let's begin by looking at an overview of the company's operations.

■ Overview of Fitter Snacker

Fitter Snacker manufactures and sells two types of nutritious snack bars: NRG-A and NRG-B. NRG-**A** touts "**A**dvanced Energy:" NRG-**B** boasts "**B**ody **B**uilding Proteins." Each bar contains the following ingredients:

- Vitamins and Minerals: important vitamins and minerals

- Dry Base Mixture: oats, wheat germ, protein powder, and spices

- Wet Base Mixture: soy milk, honey, and oil

Each type of bar contains additional ingredients that differentiate the bars: NRG-A contains carob chips and raisins: NRG-B contains hazelnuts and dates.

Fitter Snacker was founded in the 1950s and has two divisions: the Wholesale Division and the Direct Sales Division. The Wholesale Division sells to middlemen who distribute the bars to small shops, vending machine operators, and health food stores. The Direct Sales Division sells directly to large grocery stores, sporting goods stores, and other large chain stores.

The two divisions differ primarily in order quantities and pricing terms. The Direct Sales Division offers customers volume discounts to encourage larger sales orders, which are more efficient to process. The Wholesale Division charges customers a lower fixed price because the orders are usually large (otherwise, the orders would be handled by the Direct Sales Division). In both divisions, customers are sent an invoice and offered a 2% discount if they pay within 10 days; otherwise, the total balance is due in 30 days (2–10/net 30).

In addition to selling snack bars under the Fitter Snacker label, the company also packages the bars in store-brand wrappers for some chain stores.

Problems with Fitter Snacker's Sales Process

The Fitter Snacker sales process is flawed. Many sales orders have some sort of problem—incorrect pricing, excessive calls to the customer for information, delays in processing orders, missed delivery dates, and more. These problems are a result of FS's three unintegrated information systems: the sales order system, the warehouse system, and the accounting system. Information in each system is shared by performing periodic file transfers (sales order system to accounting system) and by paper printout (credit status from Accounting to sales clerks). The number of transactions handled manually creates many opportunities for data entry errors. Further, the information is not available in real time, resulting in incorrect prices and credit information.

In each sales division, Fitter Snacker has four salespeople who work on the road, plus two sales clerks who work in the sales office. Salespeople work on commission and have some leeway in offering customers "discretionary discounts" to make a sale. The entire sales process involves a number of steps that require coordination between Sales, Warehouse, Accounting, and Receiving, as shown in Figure 3.1.

FIGURE 3.1   The sales process

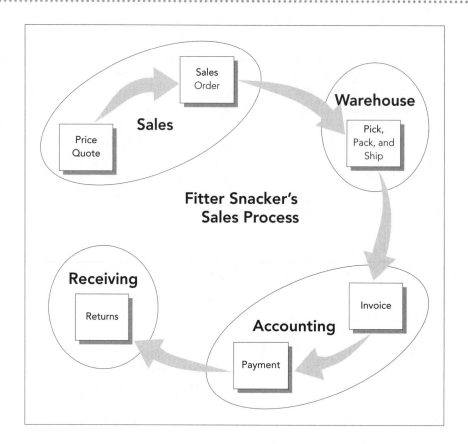

Sales Quotations and Orders

Giving a customer a price quotation and then taking the customer's order should be a straightforward process, but at FS it is not. For a new customer, the sales process begins with a sales call. At the end of the sales call, the salesperson prepares a hand-written quotation on a three-sheet, carbonless form. The top copy goes to the customer, the middle copy is faxed to the sales office, and the salesperson keeps the bottom copy for his records. The quotation form has an 800 number that the customer can call to place an order.

A number of problems can occur with this process:

■ The salesperson might make an arithmetic error in the sales quotation. For example, a salesperson in the Direct Sales Division might offer both a quantity discount plus a discretionary discount. If the salesperson isn't careful, the two discounts combined might be so deep, the company receives little or no profit.

- The salesperson faxes a copy of the sales quotation to the sales office, but sometimes the customer calls to place an order before the fax is received. The in-office clerk has no knowledge of the terms of the sale and must ask the customer to repeat the information. On the other hand, even if the quotation has been faxed, the data might not have been entered into the customer database, and still the customer must repeat the order information, much to his annoyance.

- The faxed order is a copy of a carbon copy of a hand-written quotation, and because it might not be legible, mistakes might result.

When customers place an order, they usually ask for a delivery date. To get a shipping date, the in-office clerk must contact the warehouse supervisor and ask whether the customer's order can be shipped from inventory, or whether shipping will be delayed until a future production run is delivered to the warehouse.

The warehouse supervisor can only estimate the shipping date because she's too busy to (1) make an inventory count, (2) total all orders waiting to be filled, and (3) find out how many orders are in process in the sales office.

Once the in-office sales clerk has the warehouse supervisor's estimated shipping date, she calculates how the order will be shipped and how long delivery will take. Next, the clerk checks the customer's credit status. For new customers, the clerk fills out a paper credit-check form that includes basic customer data and the amount of the order. The form is given to Accounting; they perform the credit check and then return the credit-check form with the customer's credit limit. If the credit limit is less than the amount of the purchase, the clerk calls the customer to determine what action the customer wants to take (reduce the amount of the order, prepay, or dispute the amount of credit granted). If the order is from an existing customer, then the clerk checks a paper report from Accounting that shows each customer's current balance, credit limit, and available balance. Because the report is generated weekly, it might not reflect a customer's most recent payments.

The sales clerk enters the customer's order. The computer stores the customer's order data, which are used later to analyze sales performance at the division level. It also prints out a packing list and shipping labels for the warehouse to use to pick, pack, and ship the customer's order. The computer program also produces a data file for the Accounting department to use in preparing invoices as well as for financial, tax, and managerial accounting. The data file is copied to a floppy disk and read into PeachTree, a PC-based accounting software program, on Mondays, Wednesdays, and Fridays.

Order Filling

Fitter Snacker's process for filling an order is not efficient either. Packing lists and shipping labels are printed twice a day—at noon and at the end of the day. These are carried to the warehouse, where they are hand-sorted into two groups: small orders and large orders. Fitter Snacker bars are packed 24 to a display box. The display boxes have promotional printing and are designed so they can serve as a display case. FS packs 12 display boxes together to form a standard shipping case. The warehouse stores both display boxes and shipping cases. The Production department produces, wraps, and packs the bars in display boxes. Depending on the inventory levels in the warehouse, Production personnel might transfer the display boxes directly to the warehouse, or they may pack the display boxes into shipping cases.

Display boxes and shipping cases are organized in the warehouse by label type (Fitter and store brand). Small orders consist of less than a full shipping case. For small orders, the order picker goes to the warehouse with a handcart and pulls the number of display boxes listed on the packing list. If there are not enough display boxes, the picker may break open a shipping case to get the required number of display boxes. If he does this, he is supposed to advise the warehouse supervisor so she can update her inventory records, but sometimes he forgets.

The picker then brings the display boxes back to the order packing area. Here, small orders are packed into a labeled box, with the packing list enclosed, and prepared for shipping by UPS. Large orders consist of one or more shipping cases. For these orders, the picker uses a forklift to pull the appropriate number of shipping cases to the large-order packing area. The cases are labeled for shipping, loaded on a pallet, and attached to the pallet using shrink-wrap plastic for protection. These pallets are shipped either by one of FS's two delivery trucks or by common carrier.

Inventory levels are managed by an Access database on a PC in the warehouse. Inventory levels are adjusted on a daily basis, using Production records (what has been added to the warehouse), packing lists (what has been shipped from the warehouse), and any additional sources of data (shipping cases that have been opened to pull display boxes). Monthly, there is a complete inventory count to compare the actual inventory on hand with what the inventory records show. Fitter Snacker's monthly inventory counts show that inventory records are more than 95% accurate. There are still problems filling orders, however, and these are described next.

Because snack bars are perishable, inventory levels are kept fairly low, and inventory levels change rapidly during the day. As a result, pickers might go to the shelves to pick an order and discover that there are not enough snack bars with the right label to fill the order. The picker then has to decide what to do. There may be more of that type of bar in the production area, waiting to be transferred to the warehouse, that can be used to complete the order.

For an important customer, the wrappers and display box on the production line might be changed to produce enough bars to complete the order. In other situations, the customer may want a partial shipment of whatever is on hand, with the rest shipped when it becomes available. Or, the customer might prefer a partial shipment of the goods on hand and cancel the balance of the order, preferring to place a new order later. Or, the customer may have enough inventory to wait until the whole order can be shipped, thus saving on delivery charges. To determine what to do in this situation, the order picker may have conversations with the warehouse supervisor, production supervisor, and sales clerks. Whatever the final decision, the warehouse supervisor has to contact the sales clerk so she can notify the customer (which, when things are busy, doesn't always happen) and the Accounting department so they can prepare the invoice.

Accounting and Invoicing

Invoicing the customer is problematic as well. Three times a week, sales clerks send the Accounting department the disk containing the sales order data for customer invoices. The Accounting department loads the data into the PeachTree program; then clerks manually make adjustments for partial shipments and any other changes that occurred during the order process. If corrections don't catch up to the invoices, making corrections usually delays the invoice. Sometimes, the completed invoice is waiting to be mailed when the warehouse notifies Accounting of a partial shipment, and a new invoice must be prepared. In any case, an invoice is eventually mailed to the customer, separate from the shipment.

Payment and Returns

Fitter Snacker's procedure for processing payments can yield frustrating results for customers. Almost all customers pay the invoice within 10 days to receive the 2% discount. Even though FS provides customers with two invoice copies, many customers don't include a copy of the invoice with their payment. Errors result if the correct customer's account isn't credited. In addition, if any errors occurred in the sales process, from the original quotation to the order being keyed into the sales order program, a customer will receive an incorrect invoice.

FS's returns processing is also flawed. Because FS's snack bars contain no preservatives, they have a relatively short shelf life. Thus, it's FS's policy to credit customer accounts for returned snack bars that have exceeded their "sell by" date. Credit is also given for any damaged or defective cases. Customers are supposed to call FS to get a returned material authorization (RMA) number to simplify the crediting process. When cases are returned to FS, the Receiving department fills out a returned material sheet, listing the returning company's name, the materials returned, and the RMA number. Many times, customers do not call for the RMA number or include it with their returned material. This makes it more difficult for the Accounting department to credit the appropriate account. Poor penmanship on the returned material sheet also creates problems for Accounting.

When accounts become past due, FS sends the customer a "dunning" letter. The letter informs the customer the account is past due and requests payment—if payment hasn't already been sent. As the account gets more delinquent, the dunning letters usually get more direct and threatening. If a customer's account is not properly credited, he may receive a dunning letter when he is, in fact, current. Or, the customer might be called about exceeding his credit limit after placing a new order. Both situations reduce the goodwill of good customers.

In the following sections, we will discuss how an ERP system can improve the sales process.

■ Sales and Distribution in ERP

An ERP system can improve the sales order process in several ways. Because ERP systems use a common database, they can minimize data entry errors and provide accurate information in real time to all users. An ERP system can also track all transactions (such as invoices, packing lists, RMA numbers, and payments) involved in the sales order.

Let's look at how SAP's R/3 system can manage the sales order process. In R/3 important transactions and events are assigned a number for record-keeping purposes. The electronic evidence of a transaction in R/3 is called a "document." That is the sense in which *document* is used in this section.

SAP R/3 Sales and Distribution software treats the sales order process as a cycle of events. There can be six events, from pre-sales activities to payment of the invoice. SAP R/3 provides a graphic (Figure 3.2) that summarizes the steps in the sales order process.

FIGURE 3.2 The sales and distribution process—a cycle of events in SAP R/3. *Courtesy of SAP*

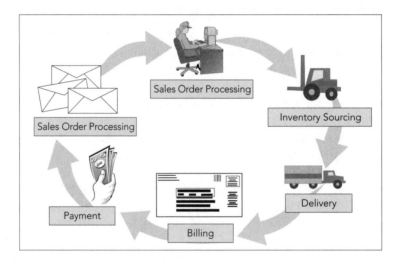

Pre-Sales Activities

The first step is called Pre-Sales Activities. At this step, a customer can get pricing information about the company's products. This information can be provided in one of two ways—an inquiry or a price quotation. Both provide pricing information to the potential customer. The difference between an inquiry and a quotation is that a written quotation is a binding document—the seller guarantees the buyer that, for some specified period of time, he can buy the product at the quoted price.

Pre-sales activities also include marketing activities such as tracking customer contacts, including sales calls, visits, and mailings. Data about customers can be maintained, so mailing lists can be generated based on specific customer characteristics, which enhances targeted marketing efforts.

Sales Order Processing

In the SAP R/3 system, Sales Order Processing is the series of activities that must take place to record a sales order. Because the R/3 system is integrated, the sales order can start from a previous quotation or inquiry. Any information that was collected from the customer to support the quotation (contact name, address, phone number) is immediately included in the sales order.

Some of the more critical steps in Sales Order Processing are recording the items to be purchased, determining the selling price, and recording the order quantities. A number of pricing alternatives are available in the R/3 system. For example, product-specific pricing, such as quantity discounts, can be configured, as well as discounts that depend on both the product and a particular customer. There is a significant amount of work required to configure a complex pricing scheme, but once the system is configured, it will automatically calculate the correct price for each customer, eliminating many problems that FS experiences.

During Sales Order Processing, the R/3 system checks the customer's credit availability by checking the Accounts Receivable tables in the R/3 database, adding the value of the order to that balance, and then comparing the result to the customer's credit limit (also available in the database). If the customer has credit available, the order is completed. If not, corrective action is taken, based on how the supplier has configured the R/3 system (reject the order, call the customer to check on recent payments, or contact Accounting).

Inventory Sourcing

When recording an order, the SAP R/3 system will check the inventory records and the production planning records to see whether the requested material can be delivered on the date the customer desires. This estimate includes the expected shipping time, taking into account weekends and holidays. Fitter Snacker's current system does not provide a good method for checking inventory availability. In the R/3 system, availability is automatically checked—and the R/3 system can be configured to increase planned production if a shortfall is expected.

Delivery

When most people hear the word *delivery*, they imagine a truck pulled up to a loading dock with cases being moved into a warehouse. Perhaps it is a result of a German-to-English translation, but in the R/3 system, the word *delivery* means releasing the documents that the warehouse will use to pick, pack, and ship orders. The delivery process allows deliveries to be created so that the warehouse and shipping activities are carried out efficiently (for example, combining like orders for picking, or grouping orders based on how and where they are to be shipped).

Once the documents are created for picking, packing, and shipping, the documents are transferred to the Materials Management module, where the warehouse activities of picking, packing, and shipping are carried out.

Billing

Next, the R/3 system uses the sales order data to create an invoice by copying the data into the invoice document. This document can be printed and mailed, faxed, or transmitted electronically to the customer. Accounting records are also updated at this point: To record the sale, Fitter Snacker's Accounts Receivable is debited (increased) and Sales is credited. Thus, R/3 updates the accounting records automatically.

Payment

When the customer sends in a payment (physically or electronically), it is again processed by the R/3 system. Cash is debited and the customer's account is credited (reduced). Notice that the timely recording of this transaction has an effect on the timeliness and accuracy of any subsequent credit checks for the customer. Fitter Snacker has a problem with getting the credit check done correctly, and this is the subject of a credit management problem in Chapter 5.

■ A Standard Order in SAP R/3

Assume that Fitter Snacker has an SAP R/3 system. Let us see how this ERP system would make FS's sales order process work more accurately and efficiently. Notice that ERP allows business processes to cut across functional area lines.

Taking an Order in SAP's R/3

Figure 3.3 shows an order entry screen in SAP R/3's Release 4.6 system. The important fields in this screen are summarized in Figure 3.4.

FIGURE 3.3 SAP R/3 order entry screen

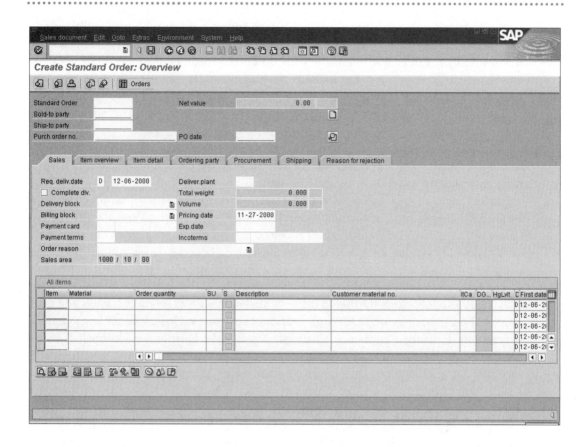

FIGURE 3.4 Data entry fields in SAP R/3

Data Entry Field	Explanation
Sold-to party	The seller's code number for the customer
Purch. order no.	The *customer's* purchase order number, which is on the customer's order form. The customer assigns this number. Note: This is not the same number that the seller assigns (via R/3) to document the sale.
Req. deliv. date	Requested delivery date—the date the customer wants to receive the order
Material	A code number for the inventory item ordered
Order quantity	The number of units the customer ordered

With R/3, users must enter code numbers for customers' names and the inventory sold, rather than using customers' names and inventory item names. Why? There could be, for example, more than one customer named "Smith Supply Company." Thus, to differentiate individual customers, a code number is assigned to each customer in the database, so code number N101 might be one Smith company, and N102 might be the other. Only one company is associated with a code number and no other. Thus, by using a code number to enter data records, there is no ambiguity about the identity of the customer. The same logic applies to distinguishing one inventory item from another. In database terminology, such codes are called "key fields."

In the R/3 system, a range of information is stored about each customer in tables called **Customer Master Data**. Master data are in the central database, and the data are used by many modules, such as SD (Sales and Distribution), FI (Financial), and CO (Controlling).

Information about materials is in tables called **Material Master Data**, which are used by the MM (Materials Management—purchasing and warehousing) and PP (Production Planning) modules. The availability of a common database to all modules is one of the advantages of having an integrated information system, because the data in the system are consistent. (This leads to problems, however, if the data are not correct, for it will be consistently incorrect in all modules.)

The R/3 system allows the user to define a number of ways to group customers and salespeople. These groupings are called **organizational structures**. One important organizational structure for FS is the Distribution Channel. With R/3, the Distribution Channel allows the user to define different ways for materials to move between the supplier and the customer. It also allows for different pricing and minimum order quantities. Defining a Wholesale Distribution Channel and a Direct Sales Distribution Channel for FS would help to ensure that customers' orders are correctly priced.

Figure 3.5 shows a completed order screen for an order from the Food City grocery chain for 190 standard shipping cases of NRG-A and 200 standard shipping cases of NRG-B.

Note that the company code for Food City is N200, and the codes for NRG-A and NRG-B are 1001 and 1002, respectively. Food City's PO number is FC1005. The PO number is a unique number provided by the customer that allows the customer to track orders with suppliers. If the customer has a question about an order, he will reference that PO number in his inquiry. Because the R/3 system records the PO number, FS can look up the status of the order using the customer's PO number.

Notice also that the name of the customer (Food City) and the names of the products (NRG-A and NRG-B) are shown, even though they were not entered directly. How did they get on the screen at this point? Once the company code is entered, the R/3 system searches the database and accesses all company information needed to complete the order. Thus, the R/3 system simplifies the data entry tasks, reducing data entry time and the possibility of error.

FIGURE 3.5 Order data inputs in SAP R/3

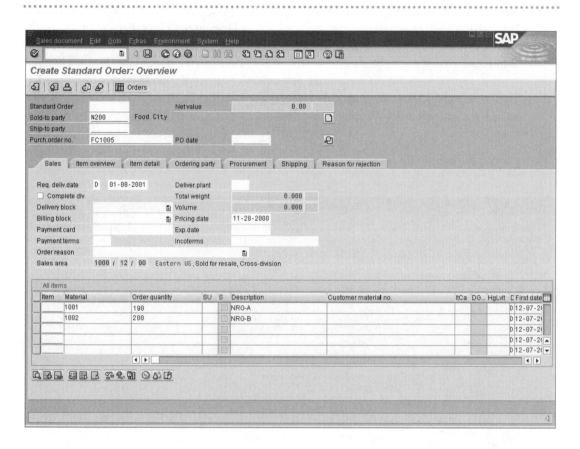

When the R/3 system saves an order, it performs inventory sourcing. Remember that when an FS customer wanted to know when an order could be delivered, the sales clerk had to go through a series of phone calls. In the R/3 system, this checking is done automatically. It includes checks on both inventory and production, and it includes shipping dates. Furthermore, the R/3 system can be configured to decide what should be done if the material for the entire order will not be available by the requested date: delay the entire order, send a partial shipment and backorder the rest, send a partial shipment and cancel the rest, cancel the order and so on. After successfully scheduling this order, SAP R/3 tells the salesperson the order has been entered into the system.

When SAP's R/3 assigns a number to a transaction, the number is called a "document number." An R/3 document number is created when a transaction is documented electronically. Later, when the goods are moved from the warehouse to be shipped, a number will be assigned to that "goods issue" and related to the sales number. Finally, when the bill is prepared for the customer, the bill's unique number (called the "invoice number") will be created and related to all the other numbers associated with the transaction.

Keeping such code numbers for transactions lets employees track the status of an order while it is in process, or they can research it after shipping. The linked set of document numbers related to an order, the "audit trail," is called a **Document Flow**. Figure 3.6 shows the Document Flow for a completed order. You can imagine that if an order includes partial shipments, partial payments, and returned material credits, the document flow can become quite complex. Without an integrated information system, the audit trail can be hard to establish, especially if many paper documents are involved. With an integrated system such as R/3, they are all linked together electronically.

FIGURE 3.6   Transaction date, or Document Flow, in SAP R/3

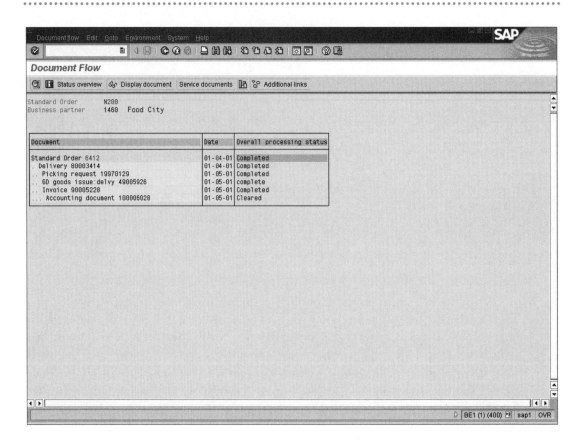

Discount Pricing in SAP R/3

When a company installs an ERP system such as SAP R/3, they can configure it for a number of pricing strategies. For example, various kinds of discounts can be allowed (per item, on all items, based on unit price, based on total value, with or without shipping charges and taxes, by individual customer, by a class of customer, and so forth). In addition, limits on the size of discounts can be established to keep salespeople from offering unprofitable discounts.

For example, suppose a salesperson wants to give a certain customer a 10% discretionary discount on NRG-B bars. The list price of the bars is $1.10 per bar, and giving a 10% price break would reduce the customer's cost to $0.99 a bar. *But* is the salesperson allowed to discount those bars? Is that discount appropriate for that customer? If the discount is allowable, will that discount be so deep that the sale will be unprofitable for FS? An ERP system automatically answers these questions.

If allowable, the SAP R/3 system would apply the discount to a specific order by changing a "pricing condition." A pricing condition should be thought of as a way to modify basic product data.

In Figure 3.7, an R/3 screen shows a discount price entry. Actions in R/3 have a condition type, here abbreviated "CnTy" in the leftmost column in the screen shown. Assume the code for a percentage discount from the list price is HA00. This code is entered into the CnTy cell, in the first available blank row, along with the discount level desired, here 10%.

FIGURE 3.7 Entering a discount off list price in SAP R/3

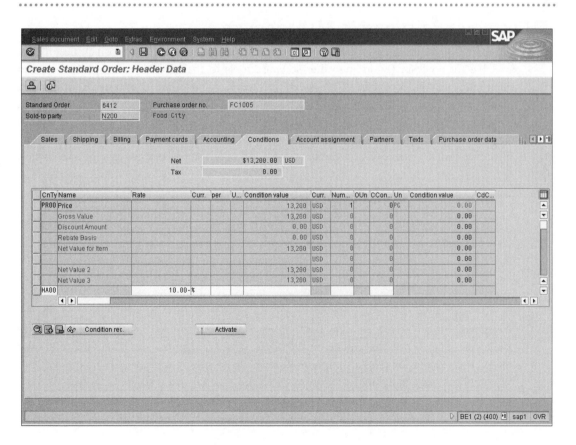

When the salesperson saves the order, SAP R/3 will check to see whether the action (giving a discount) is allowed and whether the magnitude (the percentage amount) is within limits. If the pricing conditions are satisfactory, R/3 applies them to the order. If the discount entered is not within limits, R/3 will issue an error message.

Integration of Sales and Accounting

In the current FS system, sales records are not integrated with the company's accounting records, so Accounting's billing information is not up to date.

By contrast, an ERP system would integrate sales order data with accounting data because all sales transactions are recorded in the company's central database as they are input. Thus, when Accounting retrieves data from the central database, it's retrieving the same data input by Sales, which results in up-to-date and accurate billing information. In addition, if Marketing needs a report of discounts, the database is up to date and accessible for that purpose.

Now that we've looked at how an ERP system can help a company's sales and distribution process, let's look at how data obtained during that process might be further utilized.

■ Customer Relationship Management

Companies with an ERP system have an added bonus: complete and vast quantities of data available for analysis. By adding other software tools to their ERP system, companies can extend the capabilities of their ERP system, thus increasing its value. **Customer Relationship Management (CRM)** software is one such tool.

Depending on the vendor, CRM software might be configured as a module with multiple capabilities or as a set of add-ons. (Refer to Figure 2.3 for how SAP portrays this choice.) Whatever choice is made, CRM can be viewed as a corporate strategy or as a set of activities.

CRM as a Corporate Strategy

A company uses CRM software to determine each customer's needs and then uses that knowledge to build a long-term relationship with each customer (hence Customer Relationship Management). The first step in building that relationship is to build an integrated company-wide view of each customer and then present a single "face" to each customer. Here's how it's done.

Assume that a company has many sales divisions, some of which sell to the same customers. Each division takes a somewhat narrow view of its relationship with customers, probably evaluating each relationship by the number of sales or the dollar volume of sales it makes to the customer. Top management, on the other hand, wants the big picture—the total sales and relationship profitability throughout all sales divisions. Getting this big picture requires merging data from individual divisions, which can only happen with an integrated IS, such as ERP. Thus, a company with five divisions might develop a comprehensive company-wide profile of a customer that is very different from one division's view of that same customer.

In a full CRM implementation, top management would want CRM software to do more than merge data, however. For example, when a customer contacts the company, the customer should have access to the same information about the company and its products, regardless of whether the customer telephones, e-mails, or visits the company's Web site. In addition, the customer should be able to do business—get a price quotation, place an order, track the progress of an order—through any channel. With CRM, all customer contacts are recorded in the company's database. These data, along with sales and other transaction data, allow the company to maintain a history and build a profile of the customer. This comprehensive customer information is available to all company employees who work with the customer. Thus, when the customer contacts the company, the company can use this information to meet the customer's needs in a coordinated and efficient way, paving the way to customer loyalty and future business.

CRM as a Set of Activities

CRM software exists to accomplish a variety of activities. One of the most important activities is segmenting customers. Here's how it's done.

Because ERP stores large amounts of data in one central database, important subsets of these data can be copied into a separate repository, called a **data warehouse**. A data warehouse is a database, but it's separate from a company's operational database. When users access data in the warehouse, they can analyze and manipulate data without affecting the underlying workings of their company's transaction system.

Once the data warehouse is in place, companies can use **data mining** techniques to help them sift through transaction data in the data warehouse. Companies are looking for patterns in information and relationships of data that were previously unknown. Data mining tools are sophisticated statistical and modeling software. Prior to the advent of ERP systems, companies didn't have the vast amount of information available in one database, nor did they have the sophisticated analytical and modeling tools for "mining" those data.

Thus, using data mining software, a company can stratify customers by volume, profitability, or other measures. Scarce marketing resources can then be directed to the most desirable customers, once those customers are identified; other customer relationships can be de-emphasized.

CRM software also allows these activities:

- **One-to-One Marketing**. Once a customer is categorized, products, promotions, and pricing can be tailored accordingly. Customers can be offered products related to what they are now buying ("cross-selling") or offered higher-margin products in the same lines ("upselling").

- **Sales Force Automation (SFA)**. Customer contacts are logged in the company's database. Customers that contact the company can be automatically routed to a sales representative by SFA software. SFA software can be used to forecast customer needs, based on the customer's history and transactions, and to alert sales reps accordingly. Sometimes this software is called "lead management software" because a transaction can be tracked from the initial lead to post-sale follow-up.

- **Sales Campaign Management Software.** This software lets a company organize a marketing campaign and compile its results.

- **Marketing Encyclopedias.** This software serves as a database of promotional literature about products. The material can be routed to sales reps or customers as needed.

- **Call Center Automation.** When customers call a company to get assistance with a company's products, representatives can query a knowledge management database containing information about the product. Some knowledge management software accepts queries in natural language. If novel solutions need to be developed, they can be added to the knowledge base, which thus becomes "smarter" for the next time.

All large ERP vendors offer data warehousing and data mining tools. Third-party software vendors also have written many data warehouse applications that accomplish similar tasks. No one software vendor has an end-to-end CRM software suite that meets all needs. Siebel Systems is currently a leader in many of the areas. Companies wanting many CRM functions must stitch together software from many vendors and/or write their own CRM applications. Ironically, these approaches can lead to a lack of information systems integration—exactly the problem CRM and ERP attempt to solve! To overcome this problem, ERP vendors such as SAP and PeopleSoft are busy incorporating CRM modules into their ERP programs.

Companies with existing ERP systems are more likely to use the ERP's set of data warehousing and data mining tools rather than purchase third-party software because of the ease of integration with the current system.

Why CRM Needs ERP as a Base

An ERP system provides a common transaction database. This database is used by CRM software in building customer profiles and for other CRM purposes. As a practical matter, useful subsets of transactions from the common database are taken out and put into a corporate "data warehouse." Customer contact information is then merged into the CRM data warehouse. Thus, ERP is a precursor in two ways: (1) common transaction data are used and (2) the unified database approach is retained for the company's CRM work.

The Benefits of CRM

CRM provides companies with these benefits:

- **Lower costs:** CRM can lead to operational efficiencies, such as better response times in call center operations and better use of sales force time, which lower costs.

- **Higher revenue:** By segmenting customers, better selling opportunities present themselves and revenue increases.

- **_Improved strategy and performance measurement:_** Installing and operating an ERP system requires management and staff to think of the company as a whole. This attitude carries over into CRM work. With CRM in place, management can think about different performance measures; for example, should salespeople be rewarded for exceeding sales quotas and marketing people rewarded for finding new customers? Or, should both receive rewards that are based on some measure of customer satisfaction? The former approach—typical in days before CRM and ERP—can lead to unintegrated functional thinking. The latter approach—now feasible with CRM and ERP—can lead to all personnel thinking in terms of a company-wide effort to satisfy customers.

Another Look at CRM

Customer Relationship Management is a set of related marketing-oriented activities. To give you a better grasp of CRM ideas, some real-world examples are described here.

Key customers accessing Dell Computer Corp.'s customer Web page each get a personalized page. Dell calls these pages "Premier Web Pages." The customer's page contains product and technical information that Dell feels will be important to the customer. How does Dell know this information in advance? Dell uses CRM technology to track each customer's history. Each Web page is, in effect, Dell's forecast about what the customer needs to see, based on an analysis of this history.

Dell's Web pages provide an excellent example of CRM and ERP linkages. Suppose the customer can use the Web page to order a customized computer. When a customer enters the order, it is sent to the shop floor. Data are also sent to purchasing, which lets vendors see the data for automatic reordering. The status of the order is then posted to the Web site for the customer to see.[1]

BMG Music Service's Web site acts much like Dell's site. When a member arrives at the BMG site, the company's CRM data warehouse is contacted. The customer's page is then configured to offer music that conforms to the customer's preferences.[2]

Consider another industry. Holophane Corp. makes industrial lighting fixtures, such as those seen along roadways. Lighting fixtures are custom made, so each sale is, in effect, a separate project. In the past, sales representatives maintained project specifications on paper, sharing data with customer service, product engineering, and other departments as the sales representative thought necessary. Customer service often did not have the data needed to answer a customer's questions, and those questions had to be rerouted back to the sales representative.

Holophane installed sales force automation software on its sales representatives' laptops. Order entry data are now taken by sales reps in a uniform way, uploaded to the home office, then stored in the company's database. Customer queries can now generally be answered by customer service representatives, using the data in the database.

Holophane says the software investment has been worth it, based on the company's improved sales force productivity (more sales, same staff; fewer delays answering questions). Holophane could also justify the investment by estimating benefits from customer satisfaction—increased loyalty and retention, and presumably more orders over time. Such benefits are harder to quantify than operational cost savings, but they are more in line with the need for CRM: to help establish profitable relationships with customers.[3]

Large financial firms are good candidates for productive data mining. Such firms have many customers, and some behaviors are predictable by transaction patterns. Data mining programs are often able to identify customers about to take certain actions or customers in need of certain products. For example, Royal Bank of Canada has a large data warehouse populated with data about customers' transaction history and demographics. Prepackaged queries are routinely run against the database to identify customers that might need sales help. For example, a customer's balances might be suddenly low and deposits sporadic, evidence that the customer might be thinking of leaving the bank. If the customer has been profitable to the bank—a determination that requires its own CRM calculation—a representative will call to see whether the customer has problems that need to be addressed.[4]

Phillips Petroleum used SAP's Business Information Warehouse software to help work around product shortages. Phillips had an ongoing problem with many customers placing orders at the same time, resulting in periodic product shortages and unhappy customers. SAP's Business Information Warehouse software allowed Phillips to track rail shipments and predict product shortages. Salespeople, anticipating the shortage, could contact customers and try to distribute scarce products to customers, so each customer had enough products to "tide him over" until Phillips' production could catch up with sales.[5]

Questions:

1. CRM is actually a mix of marketing and other activities. What activities are represented here?

2. Why do you think it's important to analyze data in a data warehouse rather than directly analyze the transactional database that is being accessed by the ERP system in real time?

EXERCISES

1. Identify the problems caused by Fitter Snacker's unintegrated system and then explain how an ERP system can help correct those problems. State the benefits in terms of cost and customer service that an ERP system can provide.

2. Imagine that you work for Fitter Snacker. A marketing manager says to you, "To make this proposed ERP system work, multiple sales divisions must be abandoned. A single division will need to be established, with all salespeople in the same place." How would you respond?

3. Fitter Snacker's current sales order accounting involves recording sales in each sales division and then periodically sending enough data to Accounting to record sales for the company. Complete sales order data are retained in each sales division for business analysis purposes. Assume that different divisions of the Royal Foods Company buy NRG-A and NRG-B bars from each of FS's sales divisions. To complicate matters, some divisions of Royal buy store-brand bars from FS. (Royal owns convenience store outlets.) FS management wants to see an analysis of the overall relationship with Royal. Management thinks there may be opportunities to promote their relationship with Royal, but they need to assess profitability before proceeding. They want to see what products each division sells to Royal, how much is sold, and the terms. Assume in FS's current system that all the required data are available only at the sales division level. What steps will be needed to pull this company-wide analysis together? (Review how each division sells its products and keeps its records.) Do you think a sales division manager will be enthusiastic about sharing all data with his or her counterpart in the other division? Or, do you think there might be some reluctance? Why?

4. Continuing the Royal Foods example, now assume that FS has the R/3 system installed. Each sales division records sales in the same way. Sales records are in real time and kept in the company's common database. Now what steps will be needed to pull this company-wide analysis together?

5. Draw a flow chart of the current FS sales order processing procedure, from quotation to order, to sourcing to delivery, to billing to customer payment. Assume that transactions go as expected; for example, the requested items are in stock, and there are no returns. Annotate the flow chart to show (1) when different people perform different functions, (2) the storage of data and accessing of data, and (3) when one day lapses into another as events proceed. Once you have finished drawing that flow chart, draw another flow chart, assuming that FS has the R/3 system. Once again, assume no complications. Annotate the flow chart to show (1) when different people perform different functions, (2) the storage of data and accessing of data, and (3) when one day lapses into another as events proceed.

6. With Fitter Snacker's current sales order and accounting systems, the invoice (billing) for a sale is mailed after the goods are shipped. If FS had the R/3 system, on the other hand, the invoice would be available when shipped and would actually be affixed to the shipping carton. Why does FS currently need extra time to prepare an invoice to mail after shipment? Why would this delay be avoided with R/3?

7. Fitter Snacker's NRG bars are sometimes returned when their "sell by" date expires. With an unintegrated system, the returns process is flawed in many ways, as described in this chapter. List those ways.

8. When Fitter Snacker installs R/3, their returns process is much more efficient. R/3 generates an RMA number, which the customer puts on the returning box of NRG bars. List what you think might happen when the returned goods arrive at FS. Be sure to list changes to document flow and to the original sales order.

9. Do ERP systems strengthen or weaken a corporate manager's control over operations? State your point of view and justify it with examples from this chapter.

10. Investigate uses of data mining on the Internet, and report on companies' experience with it.

11. Research articles about information systems on the Web to find out the sizes of some companies' data warehouses.

SUMMARY

1. Fitter Snacker's unintegrated information systems are at the root of an inefficient and costly sales order process. Because information is not shared in real time, customers are asked to repeat initial sales order information. As an order is processed, errors in pricing, credit checks, and invoicing also occur, presenting a poor company image to customers. Integrated ERP software would let FS avoid errors because all customer data are stored in a central database that is shared in real time by all company employees.

2. An ERP system such as SAP's R/3 sees a sale as a cycle of related functions, including taking orders, setting prices, checking product availability, checking the customer's credit line, arranging for delivery, billing the customer, and collecting payment. In R/3 all these transactions, or documents, are electronically linked, so tracking an order's status (partial shipments, returns, partial payments, and so forth) is easily accomplished.

3. When an ERP system is installed, various configuration decisions are made. These decisions reflect management's desires of how transactions should be recorded and later used for decision making. For example, the system can be configured to limit selling price discounts, thus avoiding unprofitable pricing.

4. An ERP system's central database has master data tables for customers, suppliers, and inventory. The tables hold relatively permanent information about each subject.

5. Customer Relationship Management systems build on what ERP provides, in particular on the company's common database. There are various kinds of CRM software, some available from ERP vendors and some from third-party software companies. CRM software can lead to operational savings, but most companies buy it because they feel that having better customer relationships will result in higher revenues.

■ End Notes

[1] Jennifer Bresnahan, "Improving the Odds," *CIO* (November 15, 1998) (archived at *http://www.cio.com/archive/enterprise/111509_market.html*).

[2] Bresnahan, "Improving the Odds."

[3] Meg Mitchell, "Let There Be CRM," *CIO* (March 1, 2000) (archived at *http://www.cio.com/archive/030100_let.html*).

[4] Meridith Levinson, "Slices of Lives," *CIO* (August 15, 2000) (archived at *http://www.cio.com/archive/081500_slices.html*).

[5] Beth Davis, "ERP Vendors Offer Integrated Analysis," *Information Week* (May 3, 1999), 90–92.

■ BIBLIOGRAPHY

The following articles were also used as resources:

Don Peppers and Martha Rogers, "Customer Value," *CIO* (September 15, 1998) (archived at *http://www.cio.com/archive/enterprise/091598_hs.html*).

Louise Fickel, "Know Your Customer," *CIO* (August 15, 1999) (archived at *http://www.cio.com/archive/081599_customer.html*).

Carol Hildebrand, "One to a Customer," *CIO* (October 15, 1999) (archived *http://www.cio.com/archive/enterprise/101599_customer.html*).

Lee Pender, "CRM from Scratch," *CIO* (August 15, 2000) (archived at *http://www.cio.com/archive/081500_scratch.html*).

Elana Varon, "Suite Returns," *CIO* (August 15, 2000) (archived at *http://www.cio.com/archive/08152000_suite.html*).

Mindy Blodgett, "Masters of the Customer Connection," *CIO* (August 15, 2000) (archived at *http://www.cio.com/archive/081500_overview.html*).

John Edwards, "What's Your Problem?" *CIO* (September 1, 2000) (archived at *http://www.cio.com/archive/090100_problem.html*).

Stewart Deck, "Crunch Time," *CIO* (September 15, 2000) (archived at *http://www.cio.com/archive/09152000_crunch.html*).

CHAPTER 4

Production and Materials Management Information Systems

LEARNING OBJECTIVES

After completing this chapter, you should be able to:

- Describe the steps in the production planning process of a high-volume manufacturer such as Fitter Snacker.

- Describe Fitter Snacker's production and materials management problems.

- Describe how a structured process for production and materials management planning enhances efficiency and decision making.

- Describe how production planning data in an ERP system can be shared with suppliers to increase supply-chain efficiency.

Introduction

In Chapter 2, you learned that Enterprise Resource Planning (ERP) has its roots in Materials Requirements Planning (MRP). In fact, MRP is still a large part of today's ERP systems. In this chapter, we'll look at production and materials management in an ERP system.

In Chapter 3, we looked at Fitter Snacker's sales order process, and we assumed that FS had enough warehoused snack bars to fill an order. Like most unintegrated manufacturing operations, however, FS sometimes has trouble scheduling production so its warehouse is properly stocked. In this chapter, you'll explore FS's production and materials management problems, and you'll see how ERP can help fix them.

Managers in FS's Production and Materials Management departments must answer three questions:

1. How many of each snack bar should we produce?

2. What quantities of raw materials should we order so we can meet that level of production?

3. When should we order raw materials?

An ERP system is a good tool for answering these questions because ERP has MRP capabilities and it integrates Manufacturing with other functional areas, such as Accounting. In addition, ERP can connect a manufacturer with its suppliers.

■ Production Overview

In this chapter, we will use spreadsheet examples to explain the logic that should be used to plan and schedule FS's production. After demonstrating the planning and scheduling logic at each stage of the production planning process, we will show the SAP R/3 screens that implement the spreadsheet logic. Finally, we will discuss why using an integrated information system is superior to using an unintegrated system.

The goal of production planning is to schedule production so that goods can be shipped to customers to meet promised delivery dates. There are three general approaches to production:

1. **Make-to-stock** items are made for inventory (the "stock") in anticipation of sales orders. Most consumer products (for example, cameras, canned corn, and books) are made this way.

2. **Make-to-order items** are produced to fill specific customer orders. This approach is usually taken when producing items that are too expensive to keep

in stock, or items that are made or configured to customer specifications. Examples are airplanes and large industrial equipment.

3. **Assemble-to-order** items are produced using a combination of make-to-stock and make-to-order processes. The final product is assembled for a specific order from a selection of make-to-stock components. Personal computers are a typical assemble-to-order product.

Fitter Snacker's Manufacturing Process

Fitter Snacker uses make-to-stock production to produce its snack bars. The manufacturing process is illustrated in Figure 4.1.

FIGURE 4.1   Fitter Snacker's manufacturing process

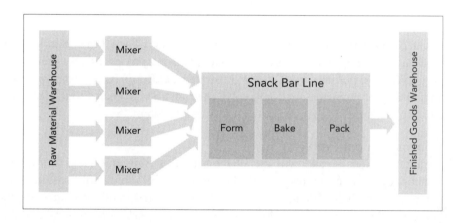

The snack bar line produces 200 bars a minute, or 3,000 pounds of bars per hour. The entire production line operates on one shift a day. Here's how it works.

Fitter Snacker's Production Sequence

Raw materials are taken from the warehouse to one of four mixers. Each mixer mixes dough in 500-pound batches. Mixing a batch of dough requires 15 minutes of mixing time, plus 15 additional minutes to unload, clean, and load the mixer for the next batch of dough; therefore, each mixer can produce two 500-pound batches of dough per hour—more than the production line can process. This extra capacity means that a mixer breakdown will not shut down the snack bar production line.

After mixing, the dough is dumped into a hopper at the beginning of the snack bar line. A forming mechanism molds the dough into bars, which will weigh four ounces after baking. Next, an automated process takes the formed bars on a conveyor belt through an oven that bakes the bars for 30 minutes. When the bars emerge from the oven, they are wrapped in a foil pouch. A group of 24 bars are packaged into a display box. At the end of the snack bar line, production personnel either stack the display boxes on pallets (for small orders) or pack the display boxes into shipping boxes and take them to the finished goods warehouse.

Changing the snack bar line from one type of bar to the other takes 30 minutes for cleaning the equipment and changing the wrappers, display boxes, and shipping cases. Each night, a second shift of employees cleans all the equipment thoroughly and sets it up for the next day's production. Thus, a changeover can be made at the end of each day without a loss of capacity. On the other hand, if two products are scheduled to be produced during one day, then a half-hour of capacity is lost in the changeover.

Fitter Snacker's Production Problems

Fitter Snacker has no problems making health food bars. Fitter Snacker has problems deciding *how many* bars to make. The manufacturing process at FS suffers from a number of problems stemming from the unintegrated nature of its information systems. Better organizational management could solve *some* of FS's problems, but many can be solved only by implementing an integrated information system.

Communications Problems

Communications breakdowns are inherent in most companies, and they are magnified in a company with an unintegrated IS. For example, FS's marketing personnel do not share information with production personnel: Marketing frequently excludes Production from meetings, doesn't consult Production when planning sales promotions, and fails to notify Production of planned promotions. Similarly, Sales does not warn Production that it has taken an exceptionally large order.

When Production must meet an unexpected increase in demand, several things happen:

- An unexpected spike in sales depletes warehouse inventories. To compensate, Production must schedule overtime production, paying for overtime labor, which results in higher production costs for products.

- Some materials such as vitamins and minerals, wrappers, and display boxes are custom products purchased from a single vendor. A sudden increase in sales demand might cause shortages or even a stockout of these materials. Getting these materials to the plant might require expedited shipping, further increasing the cost of production.

- Production personnel are evaluated on their performance. If they can't keep production costs down, they receive poor evaluations. Managers are especially frustrated when an instant need for overtime follows a period of low demand, when inventory could have been built up in anticipation of an increase in sales following a product promotion by marketing.

Inventory Problems

Week-to-week and day-to-day production planning is not linked to expected sales levels in a systematic way. When deciding how much to produce, the production manager applies rules of thumb. Her primary indicator is based on the normal amount of finished goods inventory that should be stocked and the actual inventory levels of finished goods in the warehouse. Thus, if NRG-A or NRG-B inventory levels seem low, the production manager schedules more bars for production. Her judgment is also influenced by what she hears from Marketing about expected sales.

The production manager's inventory data are in an Access database. Data records are not in real time, and they do not flag inventory that has been sold but not yet shipped. Such inventory is not available for sale of course, but this cannot be determined from the records. Thus, the level of inventory that *is* available to ship is not known. This is problematic if the Wholesale Division generates unusually large orders or a high volume of orders. For example, two large Wholesale Division orders arriving at the same time can deplete the entire available inventory of NRG-A bars. Then the Production department has to change the production schedule for NRG-B bars so it can fill the orders for NRG-A. This changeover means production of NRG-B bars must be delayed, and capacity will be lost in the changeover, creating the potential for shortages of NRG-B.

The production manager lacks not only a systematic method for meeting anticipated sales demand, but also a systematic method for adjusting production to reflect actual sales. Data on actual sales are not shared with the Production department. Hence, the production manager must use warehouse levels as a benchmark—standard levels of inventory are about the only guide available. If Production had access to sales forecasts and could adjust production levels to actual sales if sales differed significantly from expectations, inventory levels would be much closer to what is actually needed.

Accounting and Purchasing Problems

Production and Accounting do not have a good way to accumulate the day-to-day costs of production. As discussed in Chapter 3, the warehouse keeps a fairly good running count of what should be on hand. Furthermore, there is a monthly inventory taken, and actual counts are usually within 5% of what is on the books. Management would like to be more accurate, of course, but without a real-time accounting system, that goal cannot be met.

Manufacturing costs are based on the number of bars produced each day, a number that is measured at the end of the snack bar production line. Fitter Snacker uses standard costs for this purpose. For each batch of bars it produces, FS can estimate materials cost, labor cost, and related factory overhead. The number of batches is counted and multiplied by the standard cost of a batch. The resulting amount is charged to manufacturing costs.

Most manufacturing companies use standard costs in some way. The method, however, requires that standards be adjusted periodically to conform with actual costs. Actual Fitter Snacker raw material and labor costs, however, usually deviate from the standard costs. FS is not good at controlling raw materials purchases, and the production manager cannot give the purchasing manager a good production forecast. So the purchasing manager works on two tracks: (1) She tries to keep inventories high to avoid stockouts. (2) If she's offered good bulk-quantity discounts, she will buy in bulk, especially for items that have long lead-times for delivery. These purchasing practices make it difficult to forecast the volume of raw materials on hand and their average cost. Similarly, the frequent need for overtime labor means that FS has trouble accurately forecasting the average cost of labor for a batch of bars.

Thus, Production and Accounting must occasionally compare standard costs with actual costs and then adjust the accounts for the inevitable differences. This is always a tedious and unpleasant job. In theory, it would be done at each monthly closing, but often it is put off until the closing at the end of each quarter—FS's financial backers require legitimate financial statements each quarter. The adjustments are often quite large, depending on production volumes and costs during the quarter.

Exercise 4.1

1. A convenience store chain offers to buy a very large amount of its "store" brand health bars, which is the NRG-B bar with a different wrapper. The chain wants a lower-than-normal selling price, but they're also offering to place a very large order. The marketing manager asks Production, Purchasing, and Accounting whether the terms of the proposed deal will be profitable. Why will the managers in these areas have trouble providing a reliable answer on short notice?

2. The production manager notes that warehouse inventory levels are fairly high, so the production line does not need to be run for a full eight hours each day during the coming week. Workers would still need to be paid during the idle time, however, and overhead costs would be incurred as well. The production manager plans to run the line for eight hours a day anyway, because this will decrease the

average cost of bars actually produced. (Costs would be spread over more bars.) Furthermore, some warehoused raw materials will spoil if they're not used soon. Is the production manager's reasoning logical? Why or why not?

The Production Planning Process

In this section, we will go through a systematic process for production planning and materials management. The advantages of MRP and ERP will be discussed. First, we will develop spreadsheets to perform the required calculations. Then we will show the screens in the SAP R/3 system. The spreadsheets shown here are merely to simplify the presentation; MRP and ERP packages are not implemented as spreadsheets.

Production planners follow three important principles:

1. Work from a sales forecast to create an "aggregate" production plan for all products.

2. Break down the aggregate plan into more specific production plans for individual products and limited time periods.

3. Use the production plan to determine raw material requirements.

Why do planners use aggregate planning? Most companies produce many products, each of which can have many component parts. In addition, production lines might have resources—machines and workers—that are shared among products. Thus, to simplify production and planning calculations, planners tend to aggregate products into product groups. In our case, NRG-A and NRG-B will be combined and treated as snack bars. To simplify the problem, time will also be aggregated. Sales forecasts for Fitter Snacker will be calculated on a monthly basis. In turn, these monthly forecasts will then be used to determine weekly raw material orders and daily production schedules.

The R/3 Approach to Production Planning

SAP R/3's approach to the production planning process is shown in Figure 4.2. This figure will be used to trace production planning throughout this section.

FIGURE 4.2   The production planning process

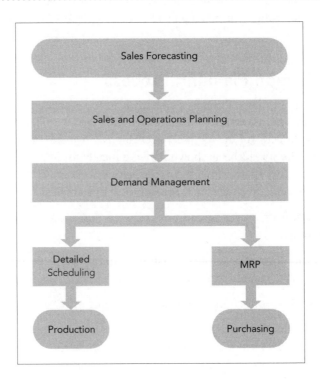

Here's an overview of the information flow at each stage:

- **Sales Forecasting** comes from a company's Marketing or Accounting department. The sales forecast represents a company's estimate of future product demand, which is the amount that customers will want to buy.

- **Sales and Operations Planning** evaluates whether the production facilities can produce enough goods to meet the consumer demand projected by the sales forecast. Conceivably, more can be sold in a period than can be made. After this analysis, an overall production plan for a time period, such as a quarter or a month, is created.

- **Demand Management** breaks down the production plan into finer time units, such as weekly or even daily production figures.

- **Detailed Scheduling** uses Demand Management's production plans as an input. Methods of detailed scheduling depend on the circumstances. For Fitter Snacker, the number of batches of each product to produce would be computed.

- **Production** uses the detailed schedule to manage its daily operations. This answers the question, "What should we be producing?"

- **MRP** determines the amount and timing of raw material orders. This answers the question, "What raw materials should we be ordering so we can meet a particular level of production, and when should we order them?"

- **Purchasing** executes purchase orders for raw materials.

In the following sections, you'll see spreadsheet examples of each stage of the production planning process shown in Figure 4.2, starting with the Sales Forecast, as shown in Figure 4.3.

Sales Forecasting

A range of forecasting techniques can be used to predict consumer demand. Fitter Snacker has no formal way of developing a sales forecast and alerting Production. SAP's R/3 system takes an integrated approach, pulling sales data from both the Sales and Distribution module and the Accounting module.

FIGURE 4.3 Sales forecasting—the first step of the production planning process

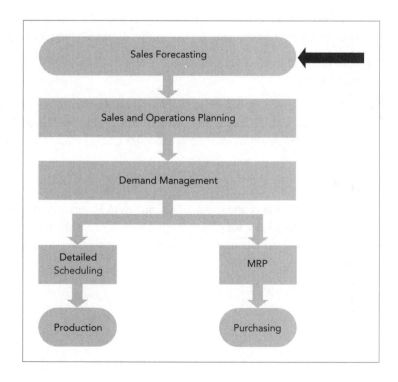

One simple technique is to use a prior period's sales and then adjust those figures for current conditions. To make a forecast for Fitter Snacker, we can use the previous year's sales data in combination with Marketing's sales projections. Look at the forecasts shown in Figure 4.4.

FIGURE 4.4   Fitter Snacker's sales forecast for January through June

Sales Forecasting		Jan.	Feb.	March	April	May	June
Previous Year (cases)		5734	5823	5884	6134	6287	6435
Growth:	3.0%	172	175	177	184	189	193
Base Projection (cases)		5906	5998	6061	6318	6476	6628
Promotion (cases)							500
Sales Forecast (cases)		5906	5998	6061	6318	6476	7128

The sales data in Figure 4.4 are for shipping cases. Each shipping case contains 12 display boxes that each contain 24 bars; thus, each shipping case contains 288 bars and weighs 72 pounds. Note the following line items in Figure 4.4:

- The forecast starts with the previous year's sales levels to reflect FS's seasonal sales fluctuations (sales are higher in the summer when more people are active).

- FS's planned sales campaign should result in a 3% growth in sales over the previous year.

- FS will be launching a special marketing promotion at the end of May to increase sales at the beginning of the summer season. FS marketing experts think this will result in an increase in sales of 500 cases for June.

Exercise 4.2

Using the data and format of the spreadsheet shown in Figure 4.4, develop a spreadsheet to forecast Fitter Snacker's sales for July through December. Make the sales growth rate of 3% be an input value, and calculate the base projection using the previous year's values shown in Figure 4.5.

Sales Volume	July	Aug.	Sept.	Oct.	Nov.	Dec.
Previous Year	6502	6327	6215	6007	5954	5813

Sales and Operations Planning

Sales and Operations Planning is the next step in the production planning process (Figure 4.6). The input to this step is the sales forecast provided by Marketing. The output of this step is a production plan designed to balance market demand with production capacity. This plan is the input to the next step, Demand Management.

FIGURE 4.6   Sales and operations planning—the second step of the production planning process

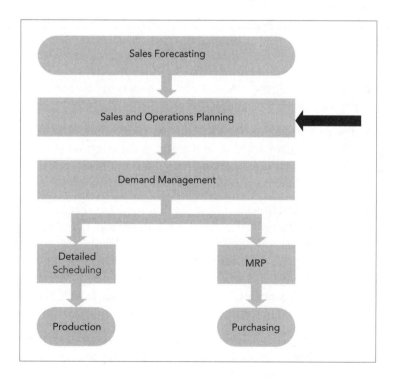

Developing a sales and operations plan from a sales forecast answers the question: "Can Manufacturing produce enough goods to meet projected sales?" In Fitter Snacker's case, there is no way to answer this question, because FS does not produce a formal estimate of sales. If FS had an ERP system, the calculation would be done as follows.

We know that Fitter Snacker can produce 200 bars per minute, so we can estimate the production capacity required by the sales forecast. Figure 4.7 shows FS's Sales and Operations Plan for the first six months of the year.

FIGURE 4.7 Fitter Snacker's sales and operations plan for June through January

Sales and Operations Planning		Jan.	Feb.	March	April	May	June
1) Sales Forecast		5906	5998	6061	6318	6476	7128
2) Production Plan		5906	5998	6061	6318	6900	6700
3) Stock Level		0	0	0	0	424	-4
4) Working Days		22	20	22	21	23	21
5) Capacity (Shipping Cases)		7333	6667	7333	7000	7667	7000
6) Utilization		81%	90%	83%	90%	90%	96%
7) NRG-A (cases)	70.0%	4134	4199	4243	4423	4830	4690
8) NRG-B (cases)	30.0%	1772	1799	1818	1895	2070	2010

- The first line is the Sales Forecast, which is the output of the Sales forecasting process shown in Figure 4.4.

- The next line is the Production Plan, a user input. The user enters different production plan amounts and observes the effect on stock levels and capacity utilization. The user continues doing this until reasonable stock and utilization levels are shown.

- The third line, Stock Level, is the difference between the Sales Forecast and the Production Plan. Notice that in May, the Production Plan is greater than the May Sales Forecast, and the Stock Level is 424. Why? Because the planners want to build up inventory to handle the increased demand in June, which is due to the normal seasonal increase in snack bar sales, plus an increase due to the planned promotional activities. The Stock Level in May is added to the Production Plan in June to meet (almost) all the Sales Forecast.

- The fourth line is Working Days, an input based on the company calendar. Using the number of working days, the available capacity each month is calculated in terms of the number of shipping cases.

200 minutes x 60 minutes x 8 hours = 96,000 bars per day,
which in terms of shipping cases is 96,000 bars/day ÷ 24 bars/box ÷ 12
boxes per case = 333.3 cases per day

So if we multiply the number of working days in a month by the production
capacity of 333.3 shipping cases per day, we will get the monthly capacity in
terms of shipping cases of snack bars, which is what is shown in line 5.

- With the available Capacity (assuming no overtime) now expressed in terms of
shipping cases, it is possible to determine the capacity utilization for each month
by dividing the Production Plan (line 2) by the available capacity (line 5),
expressing the result as a Utilization percentage (line 6). This capacity calculation
lets us see whether we have the capacity necessary to meet the production plan.

- The last step in Sales and Operations Planning is to disaggregate the production
plan. Lines 7 and 8 disaggregate the Production Plan shown in line 2, based on
the breakdown of 70% NRG-A and 30% NRG-B snack bars (which is a user
input). This monthly production plan is the output of Sales and Operations
Planning, and it will be the primary input to Demand Management.

Suppose that, based on historical data, FS is regularly able to achieve production levels at 90% of capacity. If the sales forecast requires more than 90% capacity, FS management can choose among alternatives to develop a production plan.

1. FS might choose not to meet all the forecasted sales demand, or they might
reduce promotional activities to decrease sales.

2. To increase capacity, FS might plan to use overtime production. Doing that,
however, would increase labor cost per unit.

3. Inventory levels may be built up in earlier months, when sales levels are lower, to
reduce the capacity requirements in later months. Doing that, however, would
increase inventory holding costs.

4. Management might try a hybrid approach to the capacity problem to find the
right balance: reducing sales promotions slightly, increasing production in earlier
months, and planning for some overtime production.

The production plan in Figure 4.7 creates some inventory in May to meet June's sales;
in addition, some overtime production is likely in June because capacity utilization is over
90%. This is one value of an integrated system: ERP can provide the ability to evaluate different production plans. Whereas Marketing may want to increase sales, the company
might not increase its profits if overtime costs or inventory holding costs are too high. This
sort of planning is difficult to do without an integrated information system, even for small
companies like FS. Having ERP thus helps managers keep corporate profit goals in mind.

SAP R/3 Sales and Operations Planning

In R/3 the sales forecast can be made using the Sales and Distribution (SD) module or from plans in the Controlling (CO) module. With R/3, FS's Marketing and Sales department would share data about product promotions and sales projections—information the system does not now provide. Figure 4.8 shows the Sales and Operations Planning screen from the SAP R/3 system.

FIGURE 4.8   The sales and operations planning screen in the SAP R/3 system

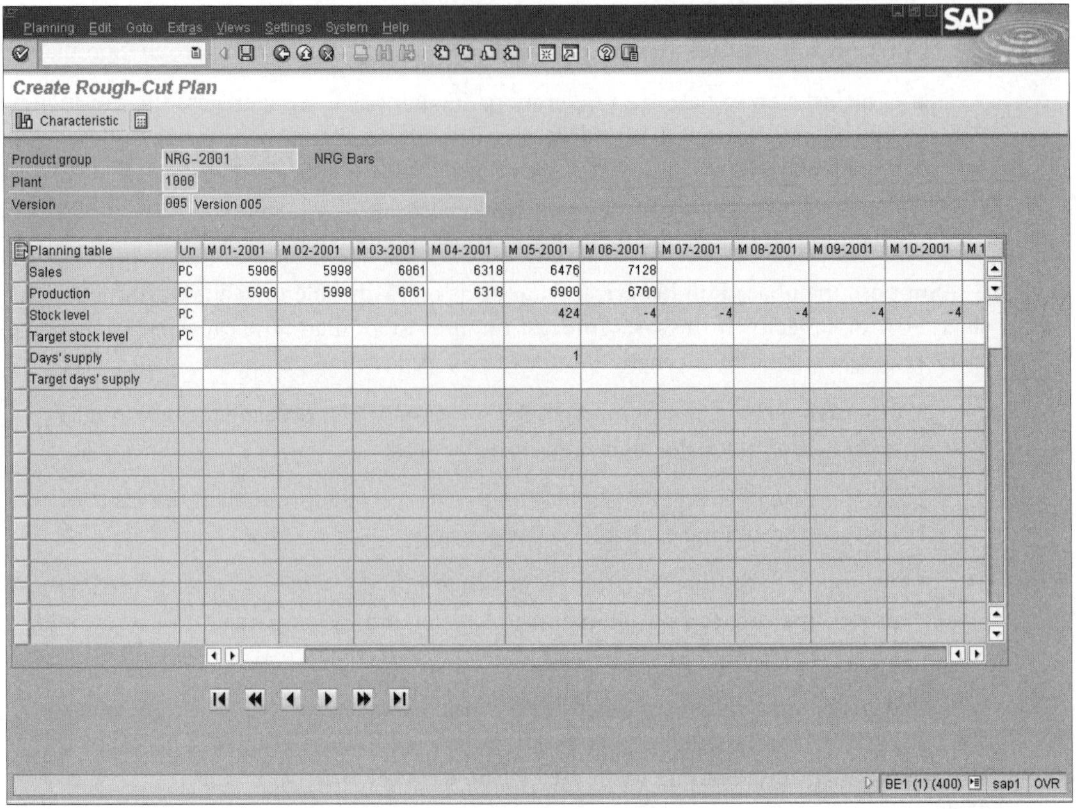

The screen allows for safety stock, which is inventory kept on hand for unexpected orders. The R/3 system maintains a set of calendars that create these calculations. The factory calendar keeps track of working days, weekends, planned holidays, and shutdowns, so that the number of actual working days is automatically available in calculations. Capacity planning is done with reference to the production plan in the R/3 system, using a variety of capacity planning methods.

Figure 4.9 shows the R/3 screen that is used to define a product group to disaggregate the Production Plan.

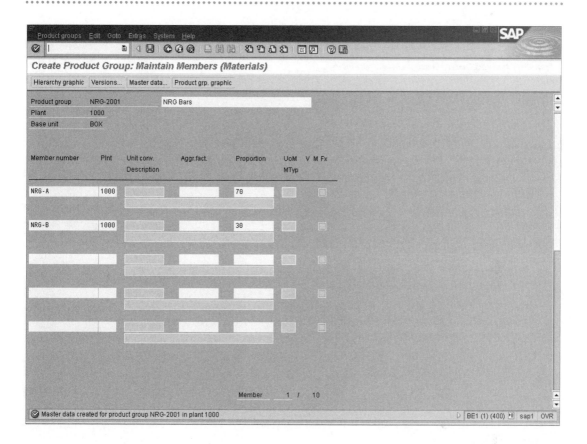

In this screen, the NRG snack bar product group consists of 70% NRG-A and 30% NRG-B. The R/3 system allows for any number of products to be assigned to a product group. In addition, product groups can be assigned to other product groups, so complicated aggregations can be defined.

Figure 4.10 shows the R/3 screen where the user transfers the Sales and Operations Plan to Demand Management. For FS, the transfer is the number of cases of each product to make in each month, given the capacity available.

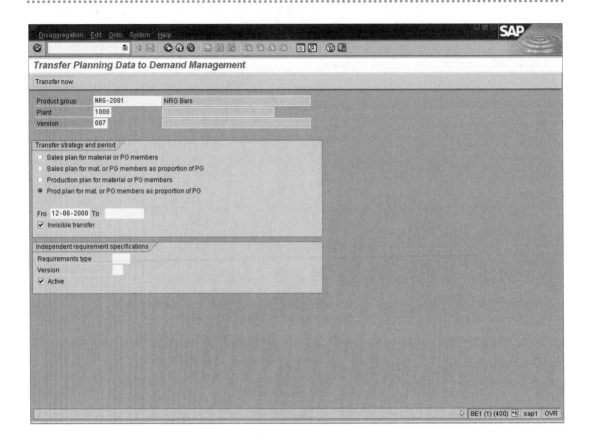

The results of the transfer in Figure 4.10 can be seen in Figure 4.11, which is the Stock/Requirements List. This screen displays the inventory level for a product, including all planned additions and reductions. As can be seen in Figure 4.11, the Production Plan from Sales and Operations Planning has been added to the Stock/Requirements List for NRG-A bars as reductions to the inventory levels.

The Stock/Requirements List is aptly named. It shows current stock, required materials, material expected to be received, and availability.

FIGURE 4.11   The Stock/Requirements List screen in the SAP R/3 system

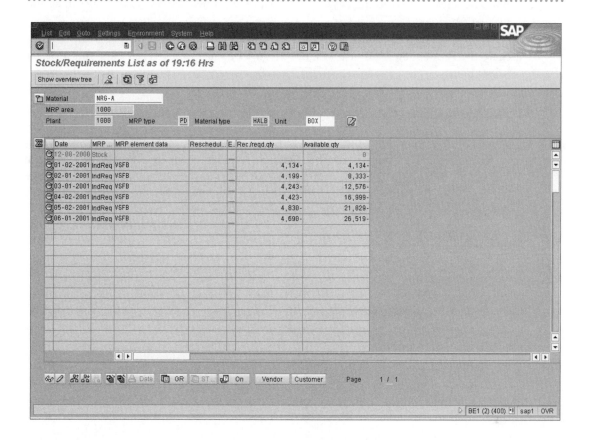

Exercise 4.3

Using the sales forecast for July through December from Exercise 4.2, develop a spreadsheet for Sales and Operations Planning. Use the format of the spreadsheet shown in Figure 4.7. The number of working days for each month is given in Figure 4.12:

FIGURE 4.12   The number of working days at Fitter Snacker, July through December

	July	Aug.	Sept.	Oct.	Nov.	Dec.
Working Days	22	18	20	23	20	17

Note that there are fewer working days in August because of an annual plant shutdown. The number of working days is also low in December because of the Christmas and New Year's holidays.

For your production plan, try to keep the capacity utilization at 95% or less. To disaggregate the Production Plan, use a figure of 70% of sales for NRG-A bars and 30% for NRG-B bars.

Demand Management

Demand Management links the Sales and Operations Planning process with Scheduling and with MRP. The output of the Demand Management process is the **Master Production Schedule (MPS)**, which is the production plan for all finished goods. For Fitter Snacker, the MPS is an input to Detailed Scheduling, which determines what to make as well as an input to the MRP process, which determines what to order.

FIGURE 4.13   Demand Management—the third step of the production planning process

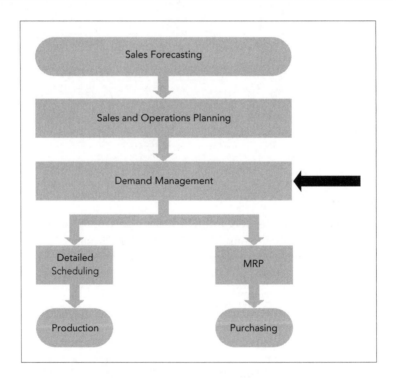

The Demand Management process splits FS's monthly production planning values into finer time periods. Figure 4.14 shows January's production plan by week and by day.

FIGURE 4.14 Fitter Snacker's production plan for January: The first five weeks of production are followed by a day-by-day disaggregation of week 1.

Demand Management		Week 1	Week 2	Week 3	Week 4	Week 5	
		1/2 - 1/5	1/8 - 1/12	1/15 - 1/19	1/22 - 1/26	1/29 - 1/31	2/1 - 2/2
Monthly Demand	NRG-A	4134	4134	4134	4134	4134	4198
	NRG-B	1772	1772	1772	1772	1772	1799
Working Days in Week		4	5	5	5	3	2
Working Days in Month		22	22	22	22	22	20
MPS	NRG-A	752	940	940	940	984	
Weekly Demand	NRG-B	322	403	403	403	422	

Demand Management		Jan 2	Jan 3	Jan 4	Jan 5	Jan 6
Monthly Demand	NRG-A	4134	4134	4134	4134	4134
	NRG-B	1772	1772	1772	1772	1772
Working Days in Month		22	22	22	22	22
MPS	NRG-A	188	188	188	188	188
Daily Demand	NRG-B	81	81	81	81	81

The weekly plan will be used for Purchasing's materials management. Daily plans will be used for scheduling the manufacturing line. The calculations were performed as follows: For the weekly plan, the production plan for NRG-A bars in week 1 was calculated as:

4,134 cases in Jan. x 4 days in week 1 ÷ 22 working days in Jan. = 751.6 cases

This was rounded to 752 cases in Figure 4.14.

Because week 5 consists of the last three days in January and the first two days in February, the production plan for NRG-A bars in week 5 was calculated as:

4,134 cases in Jan. x 3 days in week 5 ÷ 22 working days in Jan. = 563.7 cases
<u>4,198 cases in Feb. x 2 days in week 5 ÷ 20 working days in Feb. = 419.8 cases</u>
Total = 983.5 cases

Total cases were rounded to 984. In Figure 4.14, the daily production plan values for NRG-A and NRG-B bars were calculated by taking the monthly production plan value and dividing it by the number of working days in the month.

Notice that the Demand Management function for Fitter Snacker involves no user input. Input is from Sales and Operations Planning. The software also uses information from the factory calendar (working days in week and month) to calculate the output, which is the MPS. Notice also that this kind of complex planning requires software, even in small companies.

Fitter Snacker does not do this sort of planning because they have no way to formally share sales forecast data between Marketing and Production. Thus, FS cannot relate its possible sales to its capacity and to the time available to make the product. Thus, they cannot produce an accurate MPS.

The MPS is an input to Detailed Scheduling and MRP. MRP will be discussed next, and Detailed Scheduling will be briefly discussed after that.

Exercise 4.4

Develop spreadsheets like those in Figure 4.14 for July. For the weekly sales periods, the last week will include three days in August. The factory calendar information is shown in Figure 4.15.

FIGURE 4.15   Fitter Snacker's factory calendar for July

	Week 1	Week 2	Week 3	Week 4	Week 5	
Demand Management	7/2 - 7/6	7/9 - 7/13	7/16 - 7/20	7/23 - 7/27	7/30 - 7/31	8/1 - 8/3
Working Days in Week	4	5	5	5	2	3
Working Days in Month	22	22	22	22	22	18

	July 2	July 3	July 4	July 5	July 6
Demand Management					
Working Days in Month	22	22	22	22	22

MRP and Purchasing/Materials Management

Manufacturing Requirements Planning (MRP) determines the quantity and timing of the production or purchase of subassemblies and raw materials needed to support the MPS. In this section, you will see examples of how Fitter Snacker could plan its raw materials purchases if it had ERP.

FIGURE 4.16   MRP—another step in the production planning process

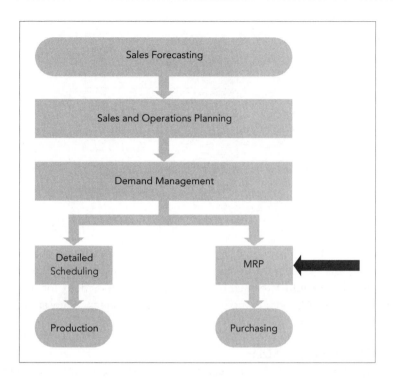

In Fitter Snacker's case, all components (ingredients, snack bar wrappers, and display boxes) are purchased, so the MRP process is used to determine the timing and quantities for purchase orders. To understand MRP, you must understand: (1) the bill of material, (2) the material's lead-time, and (3) the material's lot sizing.

Bill of Material

The **bill of material (BOM)** is the recipe, a list of the materials (including quantities) needed to make a product. (In the case of FS's snack bars, it truly is a recipe.) The recipe for a 500-pound batch of the NRG-A or NRG-B bars is shown in Figure 4.17.

FIGURE 4.17   The bill of material (BOM) for Fitter Snacker's NRG bars

	Quantity	
Ingredient	NRG-A	NRG-B
Oats (lb.)	300	250
Wheat germ (lb.)	50	50
Cinnamon (lb.)	5	5
Nutmeg (lb.)	2	2
Cloves (lb.)	1	1
Honey (gal.)	10	10
Canola Oil (gal.)	7	7
Vit./Min. Powder (lb.)	5	5
Carob Chips (lb.)	50	
Raisins (lb.)	50	
Protein Powder (lb.)		50
Hazelnuts (lb.)		30
Dates (lb.)		70

The BOM for Fitter Snacker's NRG bars is fairly simple because all ingredients are mixed together to form the dough. To produce many other products, however, component parts are joined into subassemblies that are then joined to form the finished product. Needless to say, it's more complicated to calculate the raw material requirements for products with more complex BOMs.

Lead-Times and Lot Sizing

The BOM can be used to calculate how *much* of each raw material is required to produce a finished product. To determine the *timing* and *quantity* of purchase orders, however, requires information on **lead-times** and **lot sizing**.

For example, if a manufacturer orders a make-to-stock item, the lead-time is the sum of the time required for the supplier to receive and process the order, take the material out of stock, package it, load it on a truck, and ship it to the manufacturer. In addition, the manufacturer might include the time required to receive the material in its warehouse (unloading the truck, inspecting the goods, and moving the goods into a storage location).

Lot sizing refers to the process of determining production and order quantities. In FS's case, many raw materials can only be ordered from a supplier in certain bulk quantities. For example, because FS uses large quantities of oats, the most cost-effective way to purchase oats is in bulk hopper-truck quantities, which means that the material must be ordered in 44,000-pound quantities. Wheat germ, however, is used in smaller quantities, and to avoid having wheat germ become stale, it is ordered in 2,000-pound bulk containers. Protein powder is packaged in 50-pound bags that are loaded 25 to a pallet, so the most cost-effective way to order protein powder is by the pallet load.

Let's look at the MRP process using oats, which have a two-week lead-time and must be ordered in hopper-truck quantities (multiples of 44,000 pounds). To determine when and how many oats should be ordered, we'll start with the weekly MPS for NRG-A and NRG-B bars, and then:

1. Convert the quantities from cases to 500-pound batches. This is called the "demand" for the finished product in what follows.

2. Multiply the number of batches by the pounds-per-batch quantities (which is given in the BOM) to get the gross requirements of raw material.

3. Subtract the existing raw material inventory from the demand.

4. Subtract the orders that have already been placed.

5. Plan orders in multiples of the 44,000-pound lot size, allowing for the two-week lead-time.

The five steps are summarized in the spreadsheet in Figure 4.18. This view of the data is frequently called an **MRP record**, which is the standard way of viewing the MRP process on paper.

The first two rows of the MRP record are the MPS that was the output from Demand Management. These quantities are in terms of shipping cases. The first step is to convert the MPS from shipping cases to 500-pound batches. Each shipping case weighs 72 pounds, so to convert shipping cases to 500-pound batches, multiply the number of shipping cases by 72 pounds per case, and then divide by 500 pounds per batch. Thus, producing 752 shipping cases of NRG-A bars in week 1 of the year will require 108 batches, as shown in Figure 4.18.

FIGURE 4.18   The MRP record for Fitter Snacker, weeks 1 through 5

Oats Lead-Time = 2 weeks		Week 1	Week 2	Week 3	Week 4	Week 5
MPS	NRG-A	752	940	940	940	984
(cases)	NRG-B	322	403	403	403	422
MPS	NRG-A	108	135	135	135	142
(500 lb. batches)	NRG-B	46	58	58	58	61
Gross Requirements (lbs.)		44,070	55,087	55,087	55,087	57,667
Scheduled Receipts		44,000	44,000			
Planned Receipts				88,000	44,000	44,000
On Hand	11,650	11,580	493	33,406	22,319	8,652
Planned Orders		88,000	44,000	44,000		

The next row in Figure 4.18 is called Gross Requirements. These figures are calculated by multiplying the MPS quantity (in production batches) by the pounds of oats needed for a batch of snack bars. Since FS uses 300 pounds of oats per batch of NRG-A bar and 250 pounds of oats per batch of NRG-B bar, for week 1 FS needs:

NRG-A: 108.3 batches x 300 lb. per batch   = 32,490 lb. oats
NRG-B: 46.4 batches x 250 lb. per batch   = 11,600 lb. oats
Total   = 44,090 lb. oats

Rounding makes this figure differ slightly from the value shown in Figure 4.18.

The next row in the spreadsheet is the Scheduled Receipts. This row shows the expected arrival of orders that have already been placed. There is a two-week lead-time for oats, so for oats to be available in week 1 and week 2 of the year, oats orders must be placed in the last two weeks of the previous year.

The next row, Planned Receipts, shows when Planned Orders will arrive. The Planned Receipts row is directly related to the Planned Orders row at the bottom of the record. A Planned Order is one that has not been placed with the supplier but will need to be placed to avoid running out of materials. Because there is a two-week lead-time for oat orders, the planned order will be available for production in two weeks. This is indicated by the Planned Receipts. The arrows in Figure 4.18 show the relationship between planned orders and planned receipts. For example, the planned order for 88,000 pounds of oats in week 1 will be available for use in week 3, which is shown by the planned receipt of 88,000 pounds in week 3.

The next row of Figure 4.18 is the On Hand row. The first number in this row (11,650) is the inventory of oats on hand at the beginning of week 1. The number in the week 1 column (11,580) is a projection of the inventory that will be on hand at the end of week 1 (and therefore at the beginning of week 2)—given the beginning inventory, gross requirements, and planned and scheduled receipts. In the case of the on-hand value for week 1, the initial inventory of 11,650 pounds, plus the 44,000 pounds of scheduled delivery, minus the 44,070 gross requirement leaves 11,580 pounds of oats available at the start of week 2.

The last row is the Planned Order row. This is the quantity that the MRP calculation recommends ordering, and it is the output from the MRP process that is used by Purchasing to answer the question, "What do we need to order to produce the product, and when do we need to order?"

In some cases, computerized MRP systems are configured so they will automatically place a planned order. If the MRP system is part of an ERP system, the linkage to Purchasing for this purpose would be easily made.

Many times, a planner may intervene to tell the system to adjust the planned order. For example, notice that the on-hand quantity of oats in week 2 is only 493 pounds. This means that at the start of week 2, there will only be enough oats to mix one batch of dough. Since the production line produces 3,000 pounds of bars an hour, one batch of dough will keep the production line running for only 10 minutes. If the scheduled order does not arrive early enough on the first day of week 2, the production line could be shut down. When the purchase order scheduled to arrive in week 1 was ready to be placed (two weeks prior to the beginning of week 1), the planner should have evaluated that order, considering the low inventory level projected for the beginning of week 2. The planner might have decided to place an order for two hopper-truck loads of oats, instead of the planned order for one load. Or he may have made sure that the scheduled receipt shown in week 2 would actually be delivered at the end of week 1 (Figure 4.18). Planning factors like lead-times are just estimates, and so a human being must stay in the loop to evaluate many purchase order decisions.

Notice once again the need for software to help with this kind of calculation. Of course, a human can do these computations, but with many products and constituent materials, the calculations become very tedious and are prone to error. Even for a small company such as FS, doing the calculations with reasonable speed and accuracy requires software help. Notice also the information needed to do the MRP calculation: Starting with a sales forecast, the software works down to the MPS and then to a schedule of needed raw materials.

Exercise 4.5

Develop an MRP record, similar to the one in Figure 4.18, for wheat germ for the five weeks of January. Wheat germ must be ordered in bulk container quantities, so the planned orders must be in multiples of 2,000 pounds. Use a lead-time of one week, an initial on-hand inventory of 1,184 pounds, and assume that an order of 8,000 pounds is scheduled for receipt during week 1. Are there any weeks where you, as a purchased materials planner, would consider placing an order above the minimum required? Why? Assume that there are no problems with storage capacity or shelf life.

Exercise 4.6

Fitter Snacker's purchasing policy has been to carry high levels of inventory to avoid stock-outs. With MRP-based purchasing, why can inventory levels now be lower? If you had to calculate the financial advantage of this change, how would you do it?

SAP R/3 MRP and MRP Records

Figure 4.19 shows the results of the MRP calculation process for oats.

FIGURE 4.19   The MRP list in R/3

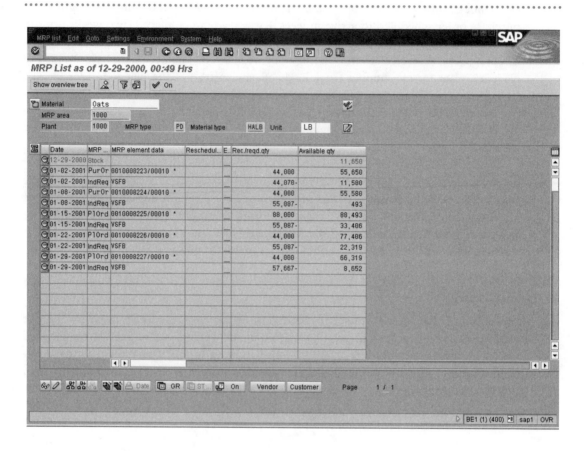

FIGURE 4.20   The Stock/Requirements List in R/3

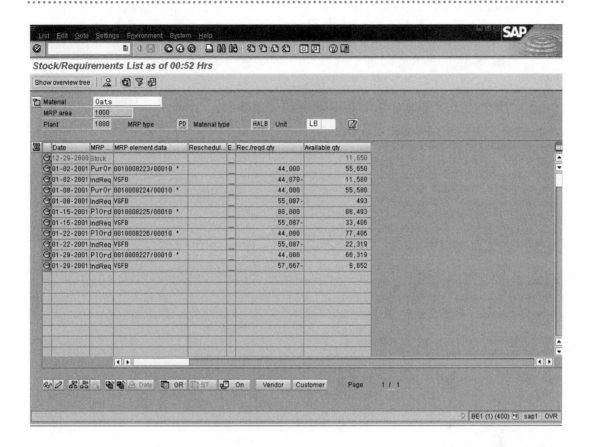

The MRP list looks very much like a Stock/Requirements List. In fact, after the MRP calculation process, it is the same (other than for some minor cosmetic differences).

The MRP list shows the results of the MRP calculations, while the Stock/Requirements List shows those results plus any changes that have occurred since the MRP list was generated (planned orders converted to purchase orders or production orders, material receipts, and so on). Because the MRP calculations are time consuming to process for a company producing hundreds of products using thousands of parts, the MRP process is usually repeated every few days—or perhaps even weekly. The Stock/Requirements List allows the users of the system to see what is happening (and will happen) with a material in real time. Compare the data shown in the MRP record in Figure 4.18 with the MRP list in Figure 4.19 and the Stock/Requirements List in Figure 4.20.

In Figures 4.19 and 4.20, SAP R/3's abbreviations are: (1) PurOr is a purchase order sent to the supplier; (2) IndReq is the demand for the product in the MRP record; and (3) PlOrd is a planned order that the system says FS should send to the supplier. Compare the data shown in the MRP record in Figure 4.18 with the MRP list in Figure 4.19 and the Stock/Requirements List in Figure 4.20. The data are the same. The Stock/Requirements List, however, will change to reflect events like receiving ordered material or a change in demand as a result of new customer orders. Once the MRP program is rerun in R/3, the two screens would be the same again. Thus, the Stock/Requirements List is in real time for the R/3 user.

Detailed Production Scheduling

Next, let's look at the last portion of the production process, Detailed Production Scheduling (Figure 4.21).

FIGURE 4.21　Detailed Scheduling—another step in the production planning process

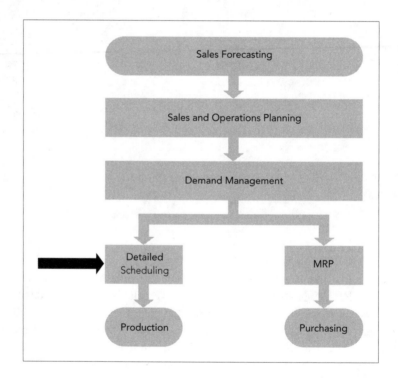

Because the capacity of the production mixers is much greater than that of the snack bar production line, scheduling mixer production is not an issue. Because the dough must be mixed before the snack bar production line can start, employees who run the mixers at FS should begin working a half-hour before the employees who run the production line. Therefore, four batches of dough are mixed before the production line starts. With a bit of a head start and a detailed schedule for the production line, it is a simple matter for the personnel operating the mixers to keep ahead of the production line. Thus, the key step is to develop a detailed production schedule for the snack bar production line.

The calculation of the detailed production schedule for the snack bar production line is somewhat complex and tedious and will not be presented in spreadsheet form, but the important concepts and issues will be discussed. A key decision in detailed production scheduling is how long to make the production runs for each product. Longer production runs mean that fewer machine setups are required, reducing the production costs and increasing the effective capacity of the equipment. On the other hand, shorter production runs can be used to lower the inventory levels for finished products. Thus, the production run length requires a balance between setup costs and holding costs to minimize total costs to the company.

In some companies, responsibility for inventory costs are assigned to a materials management group; capacity utilization performance is assigned to a production group. The materials management group wants short production runs to keep inventory levels down; the production group wants long production runs to keep capacity utilization high. Instead of being a decision in which total costs are to be minimized for the benefit of the company, it becomes a source of organizational bickering.

This points out an advantage of production planning in an ERP system. Because the goal of the company is to maximize profit, the duration of production runs should be decided by evaluating the cost of setups and holding inventory. This analysis is simplified in an integrated information system like ERP, where accounting information is collected from production and materials management transactions. This accounting information allows managers to better evaluate schedule trade-offs in terms of costs to the company.

Providing Production Data to Accounting

ERP packages do not directly connect with production machines. For example, in Fitter Snacker's case, SAP R/3 could not directly read the number of bars that came off the packing segment of the snack bar line (Figure 4.1). The data must be gathered in some way and then input into R/3 for inventory accounting purposes.

Data can be entered into R/3 via a PC on the shop floor. R/3 is an open-architecture system, meaning that automated data-collection tools marketed by third-party hardware and software companies can be integrated with R/3.

Thus, having an ERP package to record the data does not in and of itself make the shop-floor accounting data more accurate. The ERP system allows data to be entered in real time. Furthermore, capturing data for manufacturing and inventory purposes on the shop floor means that it is captured at the same time for accounting and for warehouse inventory management purposes. This eliminates any need to reconcile accounting and manufacturing records.

Exercise 4.7

Briefly describe how the implementation of SAP R/3 might change the relationship between Production and Warehousing at Fitter Snacker.

ERP in Supply-Chain Management

Manufacturing companies make finished goods by using some combination of raw material and/or subassemblies. A company must have a sufficient amount of these materials available to meet production. If too little material is on hand, not enough finished goods can be made. If too much material is on hand, material might spoil or incur extra storage costs.

Links in the Supply Chain

Manufacturers once viewed their suppliers as separate entities, but manufacturers now view their suppliers as a critical link in their manufacturing process. Supply-chain management (SCM) sees manufacturers and suppliers as part of an even larger process. Figure 4.22 provides an overview of a supply chain for manufactured goods such as Fitter Snacker's bars.

FIGURE 4.22 Supply-chain management (SCM)—from raw materials to consumer

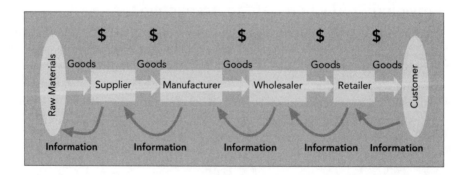

Notice that the links in the supply chain extend from the raw materials to the retail customer. For example, as part of making an automobile, iron ore is mined and transported to steel mills that produce sheet steel. Then the sheet steel goes to a fabricator, who will stamp the steel into parts that are then welded together to form the vehicle's exhaust system. The exhaust system is then transferred to an assembly line, where it is combined with thousands of other parts to manufacture the vehicle. The completed vehicle is then shipped to auto dealers and sold to consumers.

In addition to the flow of raw material from supplier to manufacturer to customer, there is a flow of money and information *back* through the supply chain. For example, when a customer buys a Fitter Snacker bar from a retailer, the retailer buys more bars from the wholesaler, who in turn buys more bars from FS. The flow of information back through the supply chain is similar.

SCM is as much of an approach to doing business with suppliers as it is a technology. A company opens its records to its suppliers, and suppliers can read the company's data because of common data formats. Working this way with suppliers cuts down on paperwork and response times. Reductions in paperwork, savings in time, and other efficiency improvements translate into cost savings for the company and the suppliers.

The Traditional Supply Chain

In a traditional supply chain, information is passed through the supply chain reactively as participants increase their product orders. For example, a retailer sees an increase in the sales of FS's bars and orders a larger quantity of bars from the wholesaler. If a number of retailers increase their orders, the wholesaler will increase its orders from Fitter Snacker. When FS gets larger orders from wholesalers, it must increase production to meet the increased demand. To increase production, FS will order more raw materials from suppliers.

Because of the time lags inherent in a traditional supply chain, it might take weeks—or even months—for information about FS's increased need for raw materials to reach FS's suppliers. Raw material suppliers may require time to increase *their* production to meet FS's larger orders, resulting in temporary shortages for the supplier.

By contrast, if the participants in the supply chain are part of an integrated process, information about the increased customer demand can be passed quickly through the supply chain, so each link in the chain can react quickly to the change.

EDI and ERP

The development of supply-chain strategies does not necessarily require an ERP system. Before ERP systems were available, companies could be linked with customers and suppliers through electronic data interchange (EDI) systems. A well-developed ERP system, however, can facilitate SCM because the needed production planning and purchasing systems are already in place. In addition, the integration of accounting data in the ERP system (described in the next chapter) allows management to evaluate changes in the market and make decisions about how those changes should affect the production plan. With an ERP system, sharing production plans along the supply chain can occur in real time. Using the Internet can make this communication faster and cheaper than if done by private EDI networks.

The Measures of Success

Performance measurements (sometimes referred to here as "**metrics**") have been developed to show the effects of better supply-chain management:

- One measure is called the **cash-to-cash cycle time**. This refers to the time from paying for raw materials to collecting cash from the customer. In one study, companies with efficient SCM processes do this in a month, whereas those without SCM take 100 days.[1]

- Another measure is total SCM costs. These costs include the cost of buying and handling inventory, processing orders, and information systems support. In one study, companies with efficient SCM processes incur costs equal to 5% of sales. By contrast, companies without SCM incur costs of up to 12% of sales.[2]

Other metrics have been developed to measure what is happening between a company and its suppliers. For example, Staples, the office-supply company, measures these facets of the relationship:[3]

- **Initial fill rate**. This is the percentage of the order that the supplier provided in the first shipment.

- **Initial-order lead-time**. This is the time needed for the supplier to fill the order.

- **On-time performance**. If the supplier agreed to requested delivery dates, this is how often the supplier actually met those dates.

Note that improvement in metrics like these would lead to improvement in overall supply chain cost measurements.

Exercise 4.8

Assume a company is considering investing in hardware and software that will improve linkages with suppliers. Management expects to save 5% of sales by tightening up the supply chain in the first year, 3% in the second year, and 1% in the third year. The company's annual sales are $1 billion. The company's chief financial officer insists that the investment must pay for itself in cost savings in three years. To meet this requirement, how much should the chief information officer be allowed to spend on developing an improved SCM system?

The Hoyt Archery Company prides itself on its exceptional products. Since the 1972 Olympics, every archer who has won a gold medal has won it using a Hoyt bow. As a result of its good reputation, the company has prospered: In 1983 the company employed 10 people; now it employs more than 150.

As the company grew, it developed two serious problems that its unintegrated system couldn't solve. To check the accuracy of their inventory records and make correcting adjustments, Hoyt would perform a complete physical inventory count twice a year. The manufacturing plant would close for three days at a cost of $5,000 per day. The plant had to close so that the inventory would not change while the count was being made.[4]

ERP also simplifies customer interaction. When a customer calls to order a bow, a customer service representative uses the Configure to Order (CTO) module to enter the order. If Hoyt has manufactured a bow of that configuration before, then the part number has already been defined and the system creates an order for it. If it is a new configuration, then the CTO module configures the bow and creates a new part number for it. New products require part numbers to be entered into the system each year. Before ERP, entering and reviewing new part numbers was a four-month task that took 10 staff members. Now three staff members can set up new part numbers in two days, and the review is completed in two weeks.[5]

Hoyt's ERP system also is allowing them to grow faster than most bow manufacturers and to enter new markets. Expansion is much smoother because the whole enterprise is integrated. Newly acquired companies' records are integrated directly into the ERP system, which is scalable (i.e., the system can grow with the company.)

Questions

1. Assume you manufacture desktop computers. They come in four different colors, with 10 types of monitors, three different speeds of CPU, and six different types of hard drives. How many different configurations are possible?

2. Now assume that the fastest CPU is sold only with the two largest hard drives. How would your answer to Question 1 change in that case?

3. Discuss why an ERP system will simplify counting inventory. Discuss why an ERP system will streamline product configuration.

4. Describe the advantage of having an ERP system in a corporate merger.

SCM is all about manufacturers linking their sales forecasts with their supplier's system. If a supplier knows exactly when a manufacturer needs materials, then the supplier can provide those materials at the right time. When a supplier can help the company to be more efficient, or vice versa, then everyone wins.

Manufacturers, suppliers, wholesalers, and retailers have been quick to embrace SCM. In fact, the most rapid rate of growth in e-commerce is from business-to-business (B-to-B) commerce, not business-to-consumer (B-to-C) commerce.

Vanity Fair Company, a supplier of uniforms, discovered that an ERP system was the key to streamlining its procurement systems.[6] Vanity Fair sells its goods to customers as large as the National Park Service and other government agencies. When those customers demanded that Vanity Fair improve its SCM, the company was quick to comply.

For starters, government customers required that Vanity Fair's goods be sold through a secure Web site. Customers needed to track and monitor orders placed with the company. In addition, the National Park Service gives each of its employees an allowance for uniforms, and it is important that allowance not be exceeded. So, the National Park Service wanted access to the sales data in Vanity Fair's information system.

Vanity Fair could comply with its customers' requests because it had an ERP system. All of Vanity Fair's data were in one place, with one numbering system, and in digital form. All divisions of the company were standardized and could electronically communicate among themselves. Tim Lambeth, VP of Global Processes at Vanity Fair, says ERP "is more critical than ever. If you are going to begin to collaborate with your suppliers, you will have to have real-time information available to them. If Wal-Mart wants to come into my system to place or track orders, it expects my system to tell it precisely what I can do and when I can do it."[7]

In fact, Wal-Mart has used SCM to gain an advantage on its competition. All customer purchases are bar-code scanned at checkout, and the data are recorded in a massive data warehouse maintained at Wal-Mart headquarters. Wal-Mart uses data mining to predict what customers are likely to buy at different times of the year. This information is shared with suppliers via an Internet link. Wal-Mart and its suppliers can then plan how products should be resupplied.

Wal-Mart is linked with 5,000 suppliers via its Retail Link program. A supplier can tap directly into the data warehouse to see how its products are selling and decide when to resupply Wal-Mart.[8]

Questions

1. Why has SCM been so important to Vanity Fair Company?

2. Why was ERP critical to Vanity Fair's SCM ?

3. Research SCM on the Web. Report on one company's success story with making SCM more efficient by using the Internet.

Another Look at Supply-Chain Management—A Supplier's Point of View

The auto manufacturing industry is highly competitive, and suppliers to that industry are required to deliver parts rapidly. In fact, when choosing a supplier, the manufacturer bases the choice not only on quality of the supplier's products, but also on the supplier's ability to deliver products rapidly. Delivering products rapidly requires supply-chain integration. Thus, all auto manufacturers require their suppliers to have computer systems that accept the manufacturers' sales forecasts.

Before e-commerce gained popularity, VDO Control Systems, a supplier of control devices and sensors, was a competitive player in the automotive industry.[9] As e-commerce gained popularity, VDO's competitors developed integrated information systems and could perform electronic acceptance of customers' orders in real time. By contrast, none of VDO's systems operated in real time. As a result of its slow information systems, VDO could not compete with the speed of its competitors' deliveries.

Lack of speed, however, wasn't VDO's only problem. Human error caused many parts to be shipped with the wrong labels, and customers returned many of those parts as defective.

VDO's cost of producing products was also higher than its competitors' costs. VDO's four manufacturing plants each had its own business processes. As a result, the manufacturing plants could not share information in a timely manner, which culminated in excess inventory and stockouts.[10]

To save the company, VDO management decided to take action, forming a team from a cross-section of the company: Manufacturing, Quality, Engineering, Finance, and IT. Using SAP's ASAP rapid implementation, the company's IS was overhauled so it could meet rapid delivery schedules and be competitive. SAP's R/3 was installed at one plant within eight months, and in the next four months the other three plants implemented the system.[11]

With the new SAP system, VDO has seen these benefits:[12]

- Sales forecasts from customers arrive online and in real time.

- Orders are automatically input in VDO's manufacturing system via a single ERP database.

- Inventory is managed company-wide, rather than plant by plant.

- Products are more easily priced.

- Parts and shipments are correctly labeled.

Questions

1. Why do you think VDO was able to implement SAP at one plant in eight months, and the other three plants four months later?

2. Why is it important to the auto manufacturers to have their sales forecasts electronically linked to their suppliers' computer systems?

◼ EXERCISES

1. Review the information at the beginning of this chapter about FS's unintegrated operation. List FS's problems, then discuss how ERP can—or can't—solve these problems.

2. Using business publications or the Internet, research a manufacturing company that has implemented an ERP system. Write a one-page report about the company, describing how the ERP system has improved the company's operations. Look especially for improvements that have come about as a result of better coordination between different functional areas (Marketing and Sales, Production, and so forth).

3. Summarize the sources and destinations of production and purchasing information within a company that has MRP and ERP. Refer to Figure 4.2 as a guide. Data could be summarized in a table with this format:

FIGURE 4.23

Data	Source	Provided to

4. Interview a production or purchasing manager at a manufacturing company. Ask the manager to describe the flow and channels of information to and from his or her department. Also ask questions: What obstacles to efficiency and productivity do you face? Do you have an MRP and/or ERP system? How does your information system help you? Hinder you? Share your findings with the class.

5. Compare customer relationship management (CRM) and supply-chain management (SCM). How are they similar? How do they differ? In answering, consider the kinds of technologies used in each.

■ SUMMARY

1. An ERP system can improve the efficiency of production and purchasing processes. Efficiency begins with Marketing sharing a sales forecast. A production plan is created based on that forecast and shared with Purchasing so raw materials can be ordered properly.

2. Production planning can be done without an ERP system, but an ERP system that contains MRP allows Production to be linked to Purchasing and Accounting. This data sharing increases a company's overall efficiency.

3. Companies are building on their ERP systems and integrated systems philosophy to practice supply-chain management. In doing this, the company looks at itself as part of a larger process that includes customers and suppliers. By using information more efficiently along the entire chain, significant cost savings can result.

■ End Notes

1 Christopher Koch, "The Big Payoff," *CIO* (October 1, 2000) (archived at *http://www.cio.com/archive/101000_payoff.html*).

2 Koch, "The Big Payoff."

3 Koch, "The Big Payoff."

4 Paula Vario, "ERP Solution Hits the Bull's Eye for Archery Manufacturer," *Midrange ERP* (March 1999): 24-26.

5 Vario, "ERP Solution Hits the Bull's Eye for Archery Manufacturer."

6 Koch, "The Big Payoff."

7 Koch, "The Big Payoff."

8 William J. Holstein, "Data Crunching Santa: Wal-Mart Knows What You Bought Last Christmas," *U.S. News & World Report* (December 21, 1998): 44.

9 Michael Stuhlreyer, "VDO Control Systems," *Midrange ERP* (January 1999): 36-38.

10 Stuhlreyer, "VDO Control Systems."

11 Stuhlreyer, "VDO Control Systems."

12 Stuhlreyer, "VDO Control Systems."

CHAPTER

Accounting and Finance

LEARNING OBJECTIVES

After completing this chapter, you should be able to:

- State and describe Accounting and Finance functional area activities.

- Identify and describe problems that arise in Accounting and Finance as a result of unintegrated information systems.

- Describe how ERP systems can solve Accounting and Finance data problems that arise in an unintegrated system.

- Explain accounting and management-reporting benefits that accrue from having an enterprise-wide information system.

Introduction

In previous chapters, you've read about functional area activities, both generally and specifically: In Chapter 2, you read an overview of functional area activities; in Chapter 3, you read about Marketing and Sales activities; and in Chapter 4, you read about Production and Materials Management. In this chapter, you'll read about the activities in a final functional area, Accounting and Finance, and how those activities are necessary for decision making.

Accounting and Finance Activities

A company's accounting system records data about the company's business transactions and related events. The data are then used for these purposes:

- Decision-makers throughout a company use accounting data to plan and manage day-to-day activities.

- Managers use accounting data to make long-range operating forecasts.

- Accountants summarize the data to generate a company's financial statements and other reports.

A company's managers will judge its accounting system by how well it helps them to make decisions and achieve company goals. Company managers want an information system that contains up-to-date data and produces accurate reports. Why? They need accurate and up-to-date information to make informed decisions, which they hope will make their company more competitive. For example, a sales manager might want a report on the profitability of his department's sales for last month. The report needs to show the products' actual sales price (list price less any discounts) as well as costs to manufacture the products. Generating this kind of report is nearly impossible unless an accounting system has timely and accurate data.

Creating Financial Statements

To create financial statements, companies must periodically summarize the balances in their accounts. This process is called **closing the books**. The financial statements include the following:

- **The balance sheet:** This statement shows account balances such as cash held, amounts owed to the company by customers, the cost of raw materials and finished-goods inventory, long-term assets such as buildings, amounts owed to vendors, amounts owed to banks and other creditors, and amounts that the owners have invested in the company. A balance sheet shows a company's financial health at the time, a key consideration for a company's creditors and owners.

- **The income statement:** This statement shows the company's sales, cost of sales, and the profit or loss for the period. Profitability is important to creditors and owners. It is also important information for managers in charge of day-to-day operations. A manager sees profits as indicators of success and losses as indicators of problems to be solved.

Many companies close the books at least quarterly, and most companies close their books more frequently. To close the books, the accounts must be checked to see that they are accurate and up to date. If a company's information systems routinely generate accurate and timely data, closing the books can go smoothly. If they are not, "adjusting" entries must be made, and closing the books can be a very time-consuming and possibly inaccurate chore.

To control a company's short-term operations, managers use sales and production forecasts, as you have seen in previous chapters. To plan long-term operations, managers use projections of sales, production, personnel needs, repayment of debt, and other factors. Expected shortfalls in funds signal a problem that Finance management must address: The company might borrow money or perhaps sell common stock. In some cases, projected shortfalls suggest deep-seated operational problems indicating that the business must change course in some way. In any case, managers expect the accounting system to provide useful forecasts for planning and operations control.

Using ERP for Accounting and Finance

Chapter 2 explains that companies have had separate functional information systems: a marketing information system, a manufacturing information system, and so on, each with its own way of gathering data and its own file system for recording data. Such unintegrated systems were built primarily to handle the needs of the functional area and secondarily to provide data to Accounting. With unintegrated systems, the functional areas shared their data with Accounting, so Accounting could keep the books. Data sharing, however, usually did not occur in real time, so Accounting's data were not up to date. In addition, the shared data might not be all that Accounting would need to prepare reports for managerial decision-makers. In that event, preparing the report would require a research effort by the accountants and functional-area clerks. Since the 1960s, legions of accountants, analysts, and programmers have been employed in an attempt to make unintegrated systems work. This approach has not worked very well, as prior chapters illustrate.

An ERP system, with its centralized database, avoids these problems. For example, suppose finished goods are transferred from the assembly line to the warehouse. The transaction can be easily recorded in the warehouse, by using a terminal or a bar code scanner to enter the data. In SAP R/3, the Materials Management module would see the event as an increase in finished-goods inventory available for shipment; the Accounting module would see the event as an increase in the monetary value of finished goods. With ERP, everyone uses the same database to record operating data. This database is then used to generate management reports, make financial statements, and create budgets.

In traditional accounting, a company's accounts are kept in a record called the "general ledger." SAP R/3 distributes control over the various general-ledger accounts among its modules. Thus, there is actually no one "Accounting" module in SAP R/3 that controls the entire accounting system. Many SAP R/3 modules cause transaction data to be entered into the database and provide views of different accounts:

- Sales and Distribution (SD) lets the user record sales.

- Production Planning (PP) causes inventory production to be scheduled and recorded.

- Materials Management (MM) controls purchasing and recording inventory.

- Financial Accounting (FI) lets the user use the general-ledger accounts to close the books and generate financial statements. FI also allows control over accounts payable and accounts receivable, presumably on the assumption that Accounting and Finance will collect money for the company and pay its bills.

- Controlling (CO) lets the user track product cost accounting.

- Human Resources (HR) lets the user generate and record the payroll.

- Asset Management (AM) lets the user record fixed-asset purchases (plant and machinery) and the related depreciation.

■ Operational Decision-Making Problem: Credit Management

A company with an unintegrated information system can have accounting data that is out of date or inaccurate. Out-of-date data can cause problems when trying to make operational decisions. This was illustrated in Chapter 3 by Fitter Snacker's problems in making credit decisions. In this section, FS's credit-granting problems will be discussed in more detail. First, we'll look at industrial credit granting in general, then at FS's credit-check problem.

Industrial Credit Management

Companies routinely sell to customers "on credit." Good financial management requires that only so much credit be extended to a customer. At some point, the customer must pay off some of the debt to justify the faith the seller has shown (and so the seller can turn his accounts receivables into cash).

In practice, sellers manage this relationship by setting a limit on how much a customer can owe at any one time and then monitoring that limit as orders come in and payments are received. For example, the seller might tell the buyer that her credit limit is $10,000, which means that the most she can owe is $10,000. If the buyer reaches that amount, no further sales orders will be accepted until the buyer pays off some of her debt. When a sale is made on credit, the seller makes an entry on the books to increase his accounts receivable and his sales. Thus, when the buyer's accounts receivable balance on the seller's books reaches $10,000, some payment must be made to the seller.

Continuing the example, assume that the buyer calls the seller to order $3,000 of goods, and her credit limit is $10,000. If the seller's accounts receivable for the buyer is *already* $8,000, then the seller should not accept the $3,000 order because the accounts receivable balance would then be $11,000, which exceeds her credit limit. In lieu of refusing the order, the seller's sales representative might suggest that the buyer reduce the size of the order—or ask the buyer to send in a payment before processing the order, thus reducing the buyer's debt. Clearly, to make this system work, a sales representative needs to be able to access an up-to-date accounts receivable balance when an order comes in.

If Marketing and Accounting have unintegrated information systems, full cooperation between the two functional areas will not be easy. Marketing knows the current order's value, but Accounting keeps the accounts receivable records. If Accounting keeps the books up to date and can provide the current accounts receivable balance to Marketing *when needed*, then credit limits can be properly managed. Marketing can compare the customer's credit limit to an accurate balance-owed amount (plus the order's value) to make a decision. However, in an unintegrated system, Accounting may not immediately record sales and/or payment receipts as they occur. In that case, accounts receivable balances will not be up to date. Furthermore, the sales clerk may be working from an out-of-date credit-balance printout. If the printout balances do not reflect recent payments, a customer may be improperly denied credit. The customer would probably challenge the denial, which would trigger a request for updated information in Accounting. The delay entailed in that research could reduce customer satisfaction; performing the research would consume valuable employee time.

These problems should not arise with an integrated information system. When a sale is made, accounts receivable is immediately increased. When a payment is received and recorded, accounts receivable is immediately decreased. Because the underlying database is available to Marketing *and* Accounting, sales representatives can also see customer records immediately. Thus, sales representatives do not need to make a request to Accounting for the customer's accounts receivable balance.

With that background, we can now consider Fitter Snacker's credit management.

FS's Credit-Management Procedures

As described in Chapter 3, an FS sales clerk refers to a weekly printout of the customer's current balance and credit limit to see if credit should be granted. Assuming the customer's order would not present credit-limit problems, the sales clerk enters the sale in the sales order-entry system, which is a stand-alone computer program. Sales data are transferred to Accounting via a floppy disk three times a week. An accounting clerk can use the sales input to prepare a customer invoice.

Accounting must make any adjustments for partial shipments before preparing the invoice. The accuracy of the adjustment process depends on whether the warehouse transmits order changes in a timely fashion. Accounting also makes the standard revenue-recognition accounting entry: debit accounts receivable and credit sales for the amount billed.

Accounting clerks also process customer payments. Checks are received and manually handled by the clerk. The data are entered in the PeachTree program, increasing cash and decreasing accounts receivable. The data are later used to make entries to individual customer accounts, reducing the amount owed to FS by customers. If time permits, accounts are posted on the day payment is received and the bank deposit made; otherwise, the entries are done as soon as possible the next day. Thus, there can be some delay from the time the customer sends a check and when the customer's accounts receivable balance is actually reduced.

Now let's look at how SAP R/3 could help FS's credit management.

Credit Management in SAP R/3

Suppose FS has SAP R/3. This system would let Marketing management set a credit limit for customers. The FS Accounting clerk would choose the Accounting module and then make a series of other menu choices to get to the credit-management part of the program. (These selections are not shown here.) Setting a credit limit for a customer means that information in the customer's Master Data file will be changed.

Assume the user is at the Credit-Management Change screen, shown in Figure 5.1, and is preparing to set Food City's credit limit.

FIGURE 5.1   SAP R/3 credit-management change screen

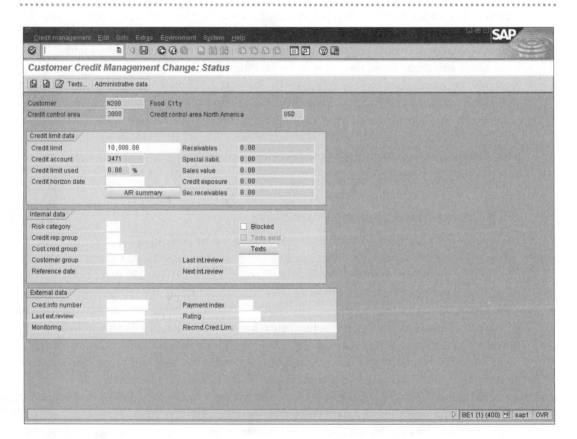

The goal is to change the credit limit value in the customer's Master Data, which contains relatively permanent customer data, such as the customer's address and credit limit. The clerk is allowed to set a credit limit for any single order and/or for the total amount owed by the customer. Suppose the clerk wants to set a total limit of $10,000 for Food City. This action is shown in Figure 5.1.

Later, a Food City order is received after the credit limit has changed. When the sales clerk saves the order, SAP R/3 automatically performs the credit check by looking up the following data in the system:

- list price of the items ordered

- accounts receivable balance

- customer credit limit

SAP R/3 computes the value of the requested order. It adds that value to the accounts receivable balance and compares the result to the credit limit. If the order exceeds the customer's credit limit, the system directs the clerk to take appropriate action. What action depends on how the system was configured—appropriate credit-management policies are precoded in the configuration. The clerk might be told to ask for payment, scale down the order, or take some other action. The important credit-management feature is this: The company will not grant credit when it should deny it, and it will not deny credit when it should be granted. (This example assumes that the order is handled by a sales representative in the home office. In Chapter 6, you will see that a field representative can access the system directly via remote means.)

Exercise

Write a short memo to your instructor in which you do the following:

1. Identify the credit-management problems caused by FS's unintegrated system.

2. Tell how the ERP system avoids those problems.

Product Profitability Analysis

Business managers use accounting data to perform profitability analyses of a company and its products. When data are inaccurate or incomplete, the analyses are flawed. But why would data be inaccurate or incomplete? There are three main reasons: (1) inconsistent record keeping, (2) inaccurate inventory-costing systems, and (3) problems consolidating data from subsidiaries. In this section, we'll look at each of those three areas.

Problem: Inconsistent Record Keeping

Each of FS's marketing divisions maintains its own records and keeps track of sales data differently: When the Direct Division records a sale, the files include a code for a sales region (Northeast, Southeast, etc.). When the Distributor Division records a sale, the files include a code for the state (New York, Delaware, etc.).

Suppose that an FS executive asks for a report that summarizes monthly sales dollars for all states in the Mid-Atlantic region for the previous year. Neither division's records would be set up to answer that question. An FS accountant would need to perform this tedious series of steps:

1. Go to source sales documents.

2. Code each by state and region (as the case may be).

3. Summarize the data by state and region to produce the report—by hand, by entering the data into a spreadsheet for review, or by some other means.

Similarly, suppose FS's management wants to evaluate the efficiency of Production's operations. Production uses paper records, so data must be taken from the paper records and entered into a spreadsheet. Many times, those paper records are inaccurate or missing, making the validity of the final report questionable.

There are many variations on this theme. Conceivably, a company's divisions do maintain the same data about a function, but often each division's systems were created at different times, and each uses a different file system. To answer a manager's question, at least one set of data would need to be rekeyed into a spreadsheet (or some other middleware program) for the merged analysis. Again, it is possible to get an answer, but doing so will take time.

Without integrated information systems, much accounting and management-reporting effort is similar to that in these examples: working around the limitations of the information systems to produce useful output.

With an ERP system, however, this sort of effort would be minimized or eliminated because both divisions would have recorded and stored their data in the same way in the same records. When the system was installed, the format of the database files would have been agreed upon as part of the system's configuration. Later, questions could be answered in a few minutes by any accountant (or manager or salesperson, for that matter) who understands how to execute a query in the database language or how to use built-in management-reporting tools.

Problem: Inaccurate Inventory-Costing Systems

Correctly calculating inventory costs is one of the most important and challenging accounting tasks in any manufacturing company. Managers need to know how much it costs to make individual products, so they can identify which products are profitable and which are not.

Inventory costing is discussed here in three sections: (1) inventory cost-accounting background, (2) how an ERP system can improve inventory cost accounting, and (3) activity-based inventory costing.

Inventory Cost-Accounting Background

A manufactured item's cost has three elements: (1) the cost of raw materials; (2) the cost of laborers working directly to produce the item; and (3) all other costs—the cost of **overhead**. Overhead costs include factory utilities, general factory labor (custodians, etc.), managers' salaries, storage, insurance, and other manufacturing-related costs.

Materials and labor are often called **direct costs** because the constituent amounts of each in a finished product can be estimated fairly accurately. On the other hand, the overhead items, called **indirect costs**, are difficult to associate with a specific product or a batch of specific products. In other words, the cause-and-effect relationship between an overhead cost (cost of heat and light) and making a particular product (NRG-A bars) is difficult to establish.

Nevertheless, overhead costs are part of making products, so some way to allocate these indirect costs to items made must be found. A common method is to use total machine hours, on the assumption that overhead is incurred to run the machines that make the products. Overhead costs are added up and then divided by total machine hours expected to get an amount per hour. This value is then used to allocate overhead costs to products. If, for example, FS's overhead per machine hour is $1,000, and 10,000 bars are made in an hour, then each bar made would be allocated ten cents of overhead ($1,000/10,000). Another allocation method distributes total overhead costs by direct labor hours, on the assumption that overhead costs are incurred so that workers can perform.

Companies (such as FS) that produce goods for inventory typically record the cost of manufacturing during a period using a **standard cost**. Standard costs for a product are established (1) by studying historical direct- and indirect-cost patterns in a company and (2) by taking into account the effects of current manufacturing changes. At the end of an accounting period, if actual costs differ from standard, adjustments to the accounts must be made to show the cost of inventory owned on the balance sheet and cost of inventory sold on the income statement.

For example, FS might determine that each NRG-A bar should cost $0.75 to make—that is, the cost of raw materials, labor, and overhead should equal $0.75, given the budgeted number of units. That would be FS's standard cost for a bar. During a month, FS might make 1 million NRG-A bars. Using the standard cost, it would increase its balance-sheet inventory account by $750,000. Also, assume that the company sells 800,000 bars in the month. In the income statement, the cost of the sales would be shown as $600,000. Also, the inventory account would be reduced by $600,000, because the company no longer has the units to sell.

If actual costs in the month equal standard costs, no balance sheet or income statement adjustments would be needed. Actual costs never exactly equal expected costs, however, so adjustments are needed. The difference between actual costs and standard costs is called **cost variances**. Note that cost variances arise with both direct and indirect costs. These variances are computed by reviewing the production records to see what costs actually occurred for products—that is, see what actually happened with costs for raw materials, direct labor, electricity, heat, forklift labor, janitorial labor, supplies, plant guards, and the many other cost elements involved with making the company's products.

If records for the various elements are kept separately, compiling variance adjustments can be quite tedious and difficult. If products are made by assembling parts that are made at different manufacturing sites, and the sites use different information systems, the adjustments may be very imprecise.

ERP and Inventory Cost Accounting

Many companies with unintegrated accounting systems infrequently analyze cost variances. Often, these companies do not know how much it actually costs to produce a unit of a product. Knowing precisely "how much it costs" can be very important. Here's why.

Suppose FS has an opportunity to sell 300,000 NRG-A bars to a new customer. This is a huge order for FS. The customer wants a price of $0.90 per bar. FS's standard cost per bar is currently $0.75, but FS knows that the costs to manufacture snack bars have been increasing significantly in the past months. FS does not want to sell at a loss per unit, but it also does not want to lose a large order or a potentially good customer. Because of the difficulty in compiling all the data to calculate cost variances, FS's cost variances are only being analyzed quarterly, and new data will not be available for another month. Should FS accept the large order?

If FS had an ERP system, costs throughout the company would have been recorded in the company-wide database as they occurred. The methods for allocating costs to products, and for computing variances, would have been built into the system when it was configured. Thus, variances could be computed automatically when needed. Not only would this simplify the process of adjusting accounts, it means that FS's management would have accurate, up-to-date information on cost variances any time. FS could decide whether it can profitably sell snack bars for $0.90 each. Furthermore, with a properly operating Sales and Operations planning process, FS could determine whether it has the capacity to complete the order on time as well! If overtime is required to complete the order, then the planning capabilities of the ERP system could be used to evaluate costs using overtime production.

ERP system configurations allow costs to be tracked using many bases—for example, by job, by work area, or by production activity. This means that unit costs can be computed using different overhead allocation bases, allowing the analyst to play "what if" with product profitability decisions. In an unintegrated system, doing such multifaceted tracking would be time consuming and difficult.

Activity-Based Costing and ERP

A trend in inventory cost accounting is toward **activity-based costing (ABC)**. In ABC, "activities" associated with overhead-cost generation are identified and then records are kept on the costs *and* on the activities. The activities are viewed as causes ("drivers") of the overhead costs. This view treats overhead costs as more direct than traditional cost-accounting methods have treated them. ABC tries to avoid rough allocation procedures in an attempt to assign costs more precisely to individual products.

Consider this example from FS's operations: Suppose that storage of raw materials is considered an "activity." Assume that storage activities differ between NRG-A and NRG-B bars because the ingredients are different, and that some of these storage activities are more labor intensive than others. In an ERP system, FS would keep track of the various activities (how often they occur) and the cost of each. Later, when determining the profitability of each kind of bar, storage costs would be added in, based on their cause: *the storage activities*. This costing is more precise than computing an average storage cost based on total storage costs and machine hours, then allocating that amount to each kind of bar. Conceivably, if the activities differ enough from one bar to the next, one could be significantly more or less profitable than the other. This fact would be revealed by the ABC approach, but *not* by traditional cost-accounting approaches. Letting managers see that difference is the value of a system that will allow ABC.

Not all overhead costs can be linked to products by their activities. However, many can, depending on the company and the manufacturing situation. For many companies, the cost and effort required to implement ABC is justified by the value of the improved information yielded.

ABC requires more bookkeeping than traditional costing methods because a company will do ABC in addition to traditional costing, and also because ABC requires keeping track of instances of activities, not just the costs. Having an integrated information system allows a company to do that much more easily. In a recent study of companies with and without ERP, it was revealed that (1) ERP companies had nearly twice as many cost-allocation bases to use in management decision-making and (2) the ERP companies' managers rated their cost-accounting system much higher.[1]

Problem: Companies with Subsidiaries

Some companies have special operations that make closing their books at the end of an accounting period a challenge. Companies that have subsidiaries or branches face such a challenge. Because company managers want "the big picture" of overall operations and profitability, account balances for each entity must be compiled and forwarded to the home office. A "consolidated" statement for the company as a whole must then be created.

You would think this would be merely an arithmetic problem: adding up cash for all the entities, accounts receivables for all the entities, and so on through the accounts. The job, however, is more difficult than that. Problems can arise from these two sources:

1. Accounts stated in another country's currency must be "translated" to U.S. dollars (in the case of a U.S. parent company).

2. Transactions between owned companies must be eliminated.

Currency Translation

Here is an example of **currency translation**. Assume the U.K. pound sterling is worth $1.60 (i.e., if you had a pound and asked a banker for its value in dollars, she would give you $1.60). A company's British subsidiary reports cash of 1 million pounds at the end of the year. When the British subsidiary's balances are consolidated with those of the U.S. parent company at the end of the year, $1,600,000 will be recorded. The same sort of "translation" would be done for all the U.K. company's accounts.

A complicating factor is that exchange rates fluctuate daily. An ERP system can be configured to access daily exchange rates and translate daily transactions automatically.

Intercompany Transactions

Here is an example of an intercompany transaction that must be eliminated. Suppose that Company A owns Company B. Company B sells raw materials to Company A for $1 million. Company A then uses the materials to make its product. Company B's sale is Company A's cost of sales. From the point of view of an outsider, money has merely passed from one part of the consolidated company to another. *A company cannot make a profit by selling to itself.* So, the effects of the intercompany sale would have to be adjusted out of the consolidation.

Companies often do business with their subsidiaries. If a company does so, there will be many such transactions. Keeping track of them and making the adjustments can be a challenge for the accountants.

Microsoft adopted SAP R/3 worldwide to handle the consolidation of its 130 subsidiaries' accounting data. A company spokesperson said:

> Prior to R/3, each entity did its own accounting on-site in separate systems, then transmitted the files to us for reloading into another system and additional manipulation. It took a huge number of people more than a week to consolidate all the numbers. With R/3, a single person does the task in a matter of hours. We had disparate systems all over the world being used for the same business functions, and the consolidation of that information was extremely costly and time consuming. The information was from all these different systems, with different field sizes, types of characters, ways of storing information, terminology, and account code block structures. We had to maintain all sorts of tables [at the home office] to map subsidiary systems to the appropriate corporate account structures, and then consolidate the information, and then reconcile any differences. This had a severe impact on getting our books closed and in providing management with timely numbers... [Microsoft adopted a single version of R/3 for all its companies]... to ensure consistency of business processes worldwide. Now every one [of our subsidiaries] is using R/3 directly to report their financial activities. At any moment in time we can look directly at financial activity around the world. Microsoft reports that the cost of the implementation will be recovered in only two years through reduced costs.[2]

Management Reporting with ERP Systems

This section is about management-reporting benefits that can be achieved with ERP. These benefits arise because of the integrated nature of the system and the use of a common database and built-in management-reporting tools.

Using Document Flow for Customer Service

As you have seen, with ERP all transactions in all areas of a company get posted in a centralized database. It is worth reemphasizing that the database *is* the company's "books." There is no separate set of books for Marketing or Production or Purchasing.

Thus, even though it is common usage to refer to "data flows" in an ERP system, it is actually a misnomer. Data do not flow from one ERP module to another because they are all in one place—the database. Each area views the same records. It might be better to speak of "data access" rather than "data flows" when talking about how these areas use the common database. Typical usage (*data flows* and *data sharing*), however, is probably too ingrained to ignore.

In SAP R/3, each posted transaction gets its own unique document number, as previously noted. This allows quick access to the data. If a user needs to look up a transaction online, he does so by referencing the document number, which acts as an index into the appropriate database-table entries.

In SAP R/3, document numbers for related transactions are associated in the database. This provides an electronic "audit trail" for analysts trying to find out what happened to an order submitted in the past. The best example of this is the linkage of document numbers for a sales order. This is the concept of "document flow" discussed in Chapter 3 and repeated in Figure 5.2.

FIGURE 5.2   Document flow of a transaction in SAP R/3

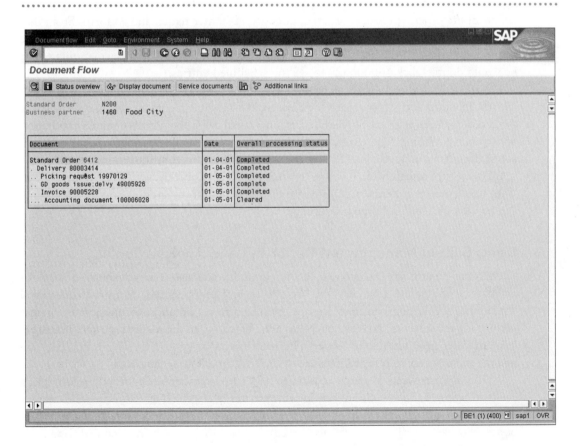

Here is a summary of the linked events shown in Figure 5.2:

- When the order was placed, document 6412 was created.

- The system recorded the Delivery. This is the transfer of the order's requirements to the Materials Management module. It is denoted by document 80003414.

- The picking request, which is the document that tells warehouse personnel what items make up the order, was created on 01-05-01 and given document number 19970129.

- The goods were removed from inventory on 01-05-01, an event recorded by document number 49005926.

- After the goods were issued, an invoice was generated so that the customer would be billed. The invoice was given document number 90005228.

- At the same time, the accounting entries for the sale were generated. The posting document number is 100006028.

The document flow can be used to "drill down" to see the details of any one of these events. For example, the user can double-click on the order number (6412) and see the details of the order—products ordered, quantity, customer name, and so forth. From that display, the user can double-click on the product numbers or the customer number to see details about them. If the user wanted to see the debits and credits in the accounting entry, he would double-click 100006028 to see the scheduled entries.

Further, the document flow can be accessed from any SAP screen. If a customer were to call and ask about the status of an order, the clerk could access the document flow and see whether the goods had been shipped. Similarly, a customer could call with an invoice number and questions about the billing. The customer-service representative could use the document flow to access documents in the chain of events, such as the original order or the picking request.

This sort of research can be done by an SAP R/3 user quickly, with a few mouse clicks. With unintegrated systems, establishing the audit trail and researching source documents can be very difficult and time consuming.

Using Built-In Management-Reporting and Analysis Tools

Accounting records are maintained in the common database. The advantage of using a database is the ability to query the records to answer prestored or ad hoc questions. Traditional accounting packages are not optimized to set up and execute queries against accounting records, but database packages are. Hence, when the records are kept in a database, the user gets double the power: The records can be kept in an accounting package *and* the records can be queried because of the built-in database language.

Thus, if a user wants to identify customer 1002's ten largest orders in the past year, a query could be executed to show the answer. In principle, this query could directly access the transaction records to get the answer. In practice, this would mean that analysts running queries would be accessing the records at the same time as current transactions are being recorded. This competition can slow down processing in even a large database, such as those used by ERP packages.

SAP's solution to this problem is to provide a "data warehouse" within each major module. Analysts query the warehouse rather than the transaction database. For example, SAP R/3 provides the Sales Information System (SIS) for analysts querying the sales records and the Logistics Information System (LIS) for analysts querying the logistics (shipping) records.

Another Look at Financial Modules—in a Service Organization

The following text is from an interview with Pat Davis, Associate Director of Management Information Systems at the University of Delaware. The University of Delaware is implementing ERP using PeopleSoft's Human Resources and Financial modules.

What benefits does the University of Delaware hope to achieve by implementing ERP in the Human Resources and Financial areas?

We were ready to upgrade our current system for both the HR and Finance areas because we had outgrown our old one. Our internal system would have required major modifications to keep up with the changing pace of our business. Also, we wanted to build into our system the "best practices" in these areas.

Our system was already integrated: We had one central database operating with three major systems on the same platform, all using the same languages. Because of the commonality, we were able to integrate all the different parts seamlessly. Also, using the Web for communications has helped in that integration. So, ERP didn't offer us integration, but what it did offer was functionality and client-server distributed computing—while maintaining central processing. For example, our current financial system couldn't manage some things unique to our business, like grants and departmental budgeting. It wasn't flexible enough. PeopleSoft's Financial module can do all the work for us without us having to program our system from scratch. So, it made sense to buy a package rather than upgrade our current system.

How did you compare vendors, and why did you choose PeopleSoft as your ERP vendor?

We asked ourselves two questions:

1. Which vendor was good in higher education?

2. Which vendor handled our important issues the best?

PeopleSoft came out on top because they offer software unique to higher education. Higher education is an unusual fit. It's not the private sector, it's not a nonprofit organization, and it's not the government. So, other vendors didn't really have a package that would work with our kind of organization. PeopleSoft has a higher-education consulting group, and yet their package still didn't fit perfectly. Higher education varies; some institutions have many grants, for example, and some have few.

Did you do a cost-benefit analysis before deciding to buy an ERP package?

Higher-education decisions are often based on criteria other than profit. There was a need to upgrade the systems, and we asked ourselves: Do we do this programming upgrade in-house, or do we buy an ERP package to do it for us? Buying the ERP package meant we would not have to upgrade our systems down the road—a substantial benefit for us. That was really our cost-benefit analysis. We didn't have a Y2K problem, so that was not an incentive to move to ERP. It seemed to us that it was faster to buy the PeopleSoft system than to upgrade the system ourselves. Plus, the PeopleSoft system had the benefit of the industry's best practices built in.

How has the implementation affected personnel at the University?

On the technical side, the majority of our MIS staff has not had any problem working with the new system. Our crew has a history of being able to move from one platform to another, so they were able to do the same here. There are some new technical skills that are required, however, for example, understanding how the package connects to our operating-system environment and our Web environment.

On the user side, it's been more difficult. [Users are University staff in areas like HR, Purchasing, and Accounting and Finance.] This implementation is really a user project. The users must work with the technical people, be they in-house or consultants, and decide how to set all the configurations for the system. Areas such as Benefits, Payroll, Purchasing, and their managers have had to decide how the PeopleSoft program should be used in their work. The users have been stretched thin during this implementation, because they still must carry on with their regular jobs.

Have you tried to customize the software?

When implementing PeopleSoft, like any other ERP package, there are hundreds of decisions to be made: Do we change the way we do the business, or do we alter the program to fit our business? You need to take your process and see how it fits with the software.

We have made some modifications along the way. The vendor supplies a proprietary language called "People Tools"; plus, add-ons can be written in C and Cobol.

What advice can you give on implementation?

One very important thing we learned the hard way was you must train your users *before* implementation. Users need to understand the system before they can start to work with the technical staff in setting all the configurations.

You must manage your consultants, or they will manage you. There are three types of consultants, and you need all three to be successful:

- *Starters*: These consultants help with targets, goals, a timeline, and the budget. A starter must be a good organizer.

- *Set-up people*: These consultants work with users in setting all the switches and configurations throughout the system.

- *Implementation and "go live" people*: These consultants are involved in the cutover to the new system.

Good communication is an important skill for the consultant. Plus, the consultant needs to know the business well. One problem for us is that we were the first university to implement PeopleSoft version 7.5, so there weren't any experienced consultants available.

Another problem was that we implemented the system in a piecemeal fashion. First, we implemented the employee system; then we moved to payroll, and last to benefits [for the HR module]. All pieces really need to be implemented simultaneously so decisions on the system can be made at the same time. With our piecemeal method, one system would set up something, and then later in another system, we would find that setup was wrong, and it would have to be changed. You need to understand the connections within the systems because there is a ripple effect.

How do you see the Web working with your new systems?

The University of Delaware was one of the first universities to go to a paperless system, using the Web for many jobs. For example, all office supplies are ordered via the Web and student grades are changed via the Web.

The newest version of PeopleSoft, version 8.0, will use XML. [XML—extensible markup language—is a step past the old Web language, HTML. XML requires the programmer to describe (tag) each piece of data. XML should become a standard language for applications that share data on the Web. Standardizing on one language will encourage smooth communication between companies on the Web.] Using XML will allow us to have a more seamless interface with the Web. Also, access into the system becomes cleaner. The 8.0 upgrade will fit nicely with our Web-oriented institution.

Questions

1. Education is a service-based industry. Identify another nonmanufacturing, service-based industry. What special ERP needs might companies in that industry have? List your ideas.

2. Using this interview as a guide, interview someone in a service-based company about their information systems needs. Report your findings to the class.

3. What do you think of Pat Davis' view that ERP systems should not be implemented piecemeal? Does this seem reasonable to you for all organizations? Why or why not?

4. You might not be familiar with how universities administer grants. At the University of Delaware, administering grants is complicated. First, the university often must commit matching funds to a grant; however, some of that money might not be spent if the grant is not funded. When grants are funded, the money can be spent in many ways. For example, the money can be spent on equipment: Any equipment costing less than $5,000 requires a grant to pay a university overhead amount, but any equipment costing less than $5,000 does not. Salary is also paid from the grant. This salary is for a researcher and perhaps graduate students. Benefits and overhead again must be deducted. Also, grant money can be paid out

as travel reimbursement or as fees to run a conference. The university needs to keep excellent records to document its expenditure on matching funds to justify to the granting agency.

a. Imagine that you are in charge of administering grants at a university. How might an ERP system help you with your job?

b. What kinds of errors might a person make who is administering grants in an unintegrated IS?

Another Look at Integrated Information Systems— in a Nonprofit Organization

ERP systems are usually thought of as systems that can promote efficiency and cost savings. In the healthcare industry, however, ERP systems can save more than money; they can also save lives.

Patients might hope that their healthcare dollars are being spent on expensive diagnostic equipment and research. Back-office ERP systems, however, can make much more of a difference in the care of some patients. Prof. Michael Smith, editor of *Healthcare Informatics Journal,* says, "Health is one of the most information-intensive businesses you will find, and that information can have a direct impact on … the quality of patient care."[3]

The typical hospital patient comes in contact with a variety of healthcare providers. Suppose a patient has had a heart attack or a stroke:

■ The patient might be seen by physicians with different medical specialties, nurses, therapists, and technicians, all of whom will record observational or test data in separate files.

■ The patient might be given a number of different drugs and treatments prescribed by different physicians.

■ The patient will be asked about his condition or lifestyle by a number of people, and these responses will be noted in different ways and in different files.

For most patients, the "information" from all those different contacts is inaccurate, disparate, or untimely:

■ Inaccurate—Physicians hand-write notes on charts, some of which are illegible. Nurses and pharmacists might misread the handwriting and take incorrect action.

■ Disparate—Sometimes, two or more physicians prescribe medication for the same patient. Dangerous drug interactions can occur when physicians don't have time to consult with each other or when a complete patient record isn't available for a doctor to read before prescribing medication. Sometimes the nurse or pharmacist filling a prescription catches the error, but not always.

- Disparate—Sometimes, two treating physicians each assume the other has prescribed the proper medications when, in fact, neither has. For example, it's common for either an internist or cardiologist to prescribe medication for a discharged patient who is recovering from a heart attack. If both assume the other is writing prescriptions and neither does, there is no system for catching the omission.

- Untimely—Scheduling tests and procedures at the proper time and in the proper sequence can be a problem with unintegrated systems. Two things are needed: (1) a list of necessary tests and procedures and when these are needed and (2) a list of equipment and services and when these are available. Matching these lists to achieve a proper schedule is very difficult when the data are in different information systems. Conceivably, a patient could be scheduled for a test or procedure when no equipment is actually available, and depending on the severity of their medical condition, this could cause drastic results.

If hospital systems were integrated, personnel could properly schedule medical tests, make sure the patient is receiving safe and adequate prescriptions, and provide one source of consistent information for all healthcare workers. In addition, streamlined procedures would allow physicians to spend more time with their patients. Currently, physicians spend an average of 25% of their time on paperwork.[4]

Systems reengineering is very challenging for organizations in the healthcare industry, not only because of technical problems but also because people resist change:

- Many different kinds of healthcare providers enter data, usually in different formats. (Physicians, for example, still usually prefer to *hand write* their entries in a paper file.)

- Different users represent different organizations—the hospital, physician's office, insurance companies, and outside care providers such as rehab centers. Each organization might have its own information-systems needs and methods.

- Users' main priority is probably not to make an information system work better. Thus, there may be resistance to changing their ways of recording data and using information systems.

In spite of the implementation challenges, U.S. healthcare organizations are spending large amounts on new IT systems. It is estimated that 6% of revenue in the U.S. healthcare industry is currently spent on IT. One likely area for improved healthcare efficiency is procurement. It is estimated that savings of 8% to 15% could be realized by improving procurement systems. Business-to-business online procurement systems could be the wave of the future for healthcare information systems reengineering.[5]

Questions

1. Suppose a patient is in a hospital that has an unintegrated information system. List five potential problems that might result from that system and how they might affect a patient's treatment.

2. Research healthcare and information systems on the Internet. Report on a healthcare organization that is implementing a new enterprise information system. What are the problems? What are the benefits of the new system?

3. Research privacy issues that impede integration of healthcare information systems. Report your findings to the class.

Another Look at Training

Now that you've seen how ERP systems work, let's take another look at training employees how to use an ERP system. As noted in Chapter 2, training is an essential part of a successful ERP implementation for the following reasons:

1. Complex ERP systems are difficult to learn and understand. It takes hours of training and practice to master a small part of the software. Learning how to use an ERP system is not like learning the basics of a spreadsheet package—it's almost like learning a whole new language.

2. One error entered into the ERP system affects all related data because all users interface with a single database. Users need to understand the business process of the entire system and how their actions impact other users.

3. To claim success, users need to actually *use* the system. Improperly trained users are likely to fear the new system and refuse to use it. There are many documented cases of failures of new systems due to lack of use.

As with most IS implementations, training begins with the project team. The training curriculum should be designed for the end user's ultimate job. Once the training materials are prepared, company employees should learn to be trainers. Internal trainers are often more effective than those outside the company because they understand the company's business processes better.[6]

Training to use an ERP system is not just learning the keystrokes; it's training the user to understand how his job fits into the larger scheme of the ERP system. A-dec, a dental equipment maker, learned this lesson the hard way. The company trained its users only how to perform their specific jobs. Users didn't understand, however, that when they entered incorrect data, they were affecting the entire system. For example, when a sales order was entered incorrectly, that also made manufacturing levels incorrect.[7]

Similarly, chemical manufacturer Solutia rushed to implement its ERP system, hoping to avoid Y2K problems. Subsequently, the company discovered its employees didn't understand the consequences of their actions. Retraining was implemented. Johnnie Foster, CIO, says the goal of the retraining was "getting people to better understand their role in the process: This is what you're doing, this is why you're doing it, and here's how it impacts other people."[8]

Hearth Technologies Inc. has a unique and successful way of training users for ERP. They purchased and equipped a double-wide trailer as a training facility. Within the trailer are PCs, printers, white boards, projection equipment, and a fax machine. The trailer is brought to the office-building area, and users can simply board the trailer for training. One advantage Hearth sees in using the "training trailer" is that users are separated from the distractions of their every-day job and can fully concentrate on learning the software. When classes are not in session, end users can go to the trailer for further practice. Jim Johnson, MIS manager at Hearth, says the trailer "is a great visual symbol as employees drive into the parking lot. It's a reminder of our commitment to ongoing education."[9]

Questions

1. Why is training so important to the success of an ERP system?

2. Enumerate the steps for successful user training in an ERP installation.

3. In what ways can ERP training improve an employee's understanding of corporate goals?

▇ EXERCISES

1. Use the Web to discover success stories that highlight advantages gained from having accounting information in an ERP system. Summarize your findings in a memorandum to your instructor. Likely sources of information are the ERP vendor sites (such as *www.sap.com*) and the site *www.erpworld.org*.

2. This exercise tests your understanding of the information needed to trace a sale through the multistep SAP R/3 sales cycle: sales order, inventory sourcing, delivery, billing, and payment. Assume that an order has been placed with your company's ERP system. These events occur:

   ■ The customer's credit is automatically checked and found to be acceptable.

   The order is recorded for the delivery date requested.

   ■ The system schedules the production of the goods. (There is not enough inventory to ship from stock.)

   ■ The system schedules raw material orders from the vendors to make the goods ordered.

   ■ The raw materials are received and stored.

   ■ The goods are produced and reserved for shipment to the customer.

   ■ The delivery is scheduled, and the goods are put on the delivery truck. An invoice is printed and included with the shipment.

   ■ Accounting is notified of the shipment's details.

- A month later, the customer sends in payment, which is recorded in accounting.

For each of these events, enumerate the information that must be recorded in the common database. You do not need to know how to use a database to do this, nor do you need to understand bookkeeping. At each step, did the wealth of the company increase or decrease? At each step, how did the company's obligations to outsiders change? At each step, how did the obligations of outsiders to the company change?

3. To answer these questions, review how sales are processed in FS's divisions. Suppose it is 4 p.m., September 30, 2001. Fitter Snacker's CEO sends this e-mail message to the accounting manager: "I need to meet with the Board of Directors tomorrow morning. The Board members are concerned about the current sales of our energy bars. They would like to see sales data for today, 9/30, as a typical day. Please complete the attached report and have it on my desk by 9 A.M. tomorrow, October 1." The blank table shown in Figure 5.3 is attached to the memo.

FIGURE 5.3

Sales data September 30	Distributor Division	Direct Sales Division	Total
# Bars sold			
NRG-A			
NRG-B			
Total			
$ Value of bars sold			
NRG-A			
NRG-B			
Total			
# Customers sold to			
NRG-A			
NRG-B			
Total			

Given the company's sales-order data-processing practices, why won't the accountants be able to answer this question easily? Cite the practices that cause difficulty and explain. If FS had an ERP system, why would this question be easy to answer? To answer this question, review Chapters 3 and 4 to see how sales orders are processed.

4. The following exercise will test your understanding of FS's current credit-check system. In each situation, you are given background data and information about documents in the system.

SITUATION 1

Background data:

FIGURE 5.4

Today's date	6/29/00
Current list price, NRG-A	$1.00/bar
Current list price, NRG-B	$1.10/bar
Accounts receivable balance at start of business day, ABC Corp.	$8,000
Credit limit, ABC Corp.	$10,000

Current order:

FIGURE 5.5

Product	NRG-A
Amount	4 cases (1152 bars)
Price	List
Ship to	ABC headquarters
Date desired	7/5/00
Next invoice number	A1001

Documents in system:

FIGURE 5.6

> No documents relating to ABC are in the system.

a. Given the state of the system, will credit be granted or denied for ABC Corporation's current order?

b. If the system processed data in a more timely way, should credit be granted or denied?

SITUATION 2

Background data:

FIGURE 5.7

Today's date	7/3/00
Current list price, NRG-A	$1.00/bar
Current list price, NRG-B	$1.10/bar
Accounts receivable balance at start of business day, KLM Corp.	$5,000
Credit limit, KLM Corp.	$7,000

Current order:

FIGURE 5.8

Product	NRG-B
Amount	5 cases (1440 bars)
Price	List
Ship to	KLM headquarters
Date desired	7/8/00
Next invoice number	A1200

Documents in system:

FIGURE 5.9

Purchase order KLM 82332 for three cases (864 bars) of NRG-A. This order is in the sales order-entry program, but it has not been transferred to the PeachTree program (thus, Accounting does not yet know about this sale).

c. Given the state of the system, will credit be granted or denied for KLM Corporation's current order?

d. If the system processed data in a more timely way, should credit be granted or denied?

SITUATION 3

Background data:

FIGURE 5.10

Today's date	7/4/00
Current list price, NRG-A	$1.00/bar
Current list price, NRG-B	$1.10/bar
Accounts receivable balance at start of business day, ACORN Corp.	$5,000
Credit limit, ACORN Corp.	$5,000

Current order:

FIGURE 5.11

Product	NRG-A
Amount	150 boxes (3,600 bars)
Price	List
Ship to	ACORN headquarters
Date desired	7/6/00
Next invoice number	A1300

Documents in system:

FIGURE 5.12

· ·

> A check from ACORN was received in yesterday's mail and entered into the PeachTree Accounting system. The check is for $2,000, applied to invoices from June's sales. The sales clerks are working from credit-limit printouts prepared at the beginning of the week (three days ago).

e. Given the state of the system, will credit be granted or denied for ACORN Corporation's current order?

f. If the system processed data in a more timely way, should credit be granted or denied?

5. ERP systems save time for accountants in many ways. From reading this chapter and other sources with which you are familiar, list those ways and provide examples.

6. Companies usually prepare division and company-wide budgets each year. These budgets show important monthly results that the company plans to achieve: sales, cost of sales, inventory levels, cash on hand, and other key data. Such budgets are effective for planning and controlling operations if they are designed as "flexible" budgets. With a flexible budget, the budget is restated as conditions change from month to month, so goals can be reasonably achieved and used for evaluating performance. If information systems are unintegrated, getting data from departments within the company to create the budget is a chore. Keeping the data up to date is very difficult and often not done. If not kept up to date, budgets are not useful for planning and controlling operations. Thus, the "flexible budget" concept is a good idea, but it is difficult to achieve. ERP makes flexible budgeting more achievable, and so ERP helps management discharge its planning and controlling roles better. Why do you think flexible budgeting would be more achievable with ERP systems? List your reasons and explain.

7. Review (in Chapter 4) FS's unintegrated production and purchasing procedures. How would its current job-scheduling, production, and purchasing procedures result in variances from standard costs? Why would FS have trouble researching these costs at month's end to adjust the "standard costs per unit" to accurate "actual costs per unit"?

■ SUMMARY

1. A company needs an accounting system to record transactions and generate financial statements. The accounting system should let the user summarize data in meaningful ways. The data can then be used to assist decision-makers in their day-to-day work and also in long-range planning.

2. With unintegrated information systems, accounting data might not be up to date, and this can cause problems when trying to make operational decisions, such as granting credit. Data can also be inaccurate because of weaknesses in unintegrated systems, and this problem can have an effect on decision-making and profitability.

3. Closing the books at the end of an accounting period can be difficult with an unintegrated IS, but it is relatively easy with an integrated IS.

4. Using an integrated IS and a common database to record accounting data has important inventory cost-accounting benefits. More precise record keeping is possible, and this can lead to more accurate product cost calculations. These, in turn, can help managers decide which products are profitable and which are not.

5. The use of an integrated system and a common database to record accounting data has important management-reporting benefits. The user has built-in drill-down and query tools available as a result.

■ End Notes

1. Kip R. Krumweide and Win G. Jordan, "Reaping the Promise of Enterprise Resource Systems," *Strategic Finance* (October 2000): 49–52.
2. "Success Story: Microsoft," online and previously archived at *http://www.erpworld.org/erp4a.html.*
3. Fiona Harvey, "Inside Track: Information at the Bedside: Healthcare Computer Systems: Technology Is Making Healthcare Administration More Efficient," *Financial Times* (online) (November 28, 2000).
4. Harvey, "Inside Track: Information at the Bedside."
5. Harvey, "Inside Track: Information at the Bedside."
6. Katherine Scannell, "Education and Training: The Key to ERP Success," *Midrange ERP* (February 1999): 6–12.
7. Malcolm Wheatley, "ERP Training Stinks," *CIO* (online), June 1, 2000.
8. Wheatley, "ERP Training Stinks."
9. Louise Keppel, "Hearth Technologies' Training Trailer," *Midrange ERP* (February 1999): 8.

CHAPTER 6

ERP and the World Wide Web

LEARNING OBJECTIVES

After completing this chapter, you should be able to:

- Name and describe the types of e-commerce.

- State why ERP is essential to the success of a company engaged in e-commerce.

- Describe what an ASP does.

- State how ERP is delivered to users by an ASP.

- Provide examples of wireless connectivity to ERP systems and explain why they are important.

Introduction

This chapter is about how companies are using Enterprise Resource Planning (ERP) systems with the World Wide Web. As you've read, an ERP system lets a company accomplish things that cannot be done well, if at all, without ERP. In this chapter, you'll learn how effectively competing in high-volume e-commerce may be impossible without the infrastructure provided by ERP. In addition, you will learn how companies can use the Web and "rent" ERP software from special-purpose software companies.

Also in this chapter, you'll read about connectivity. Original ERP designs usually assumed that a certain number of users would access the system from designated computers. In the past few years, mobile workers have been asking for remote access to companies' internal systems. ERP vendors have worked hard to provide Internet and wireless connectivity to their systems.

■ E-Commerce Background

In the not-too-distant future, most companies will conduct at least part of their business operations through **e-commerce**, that is, on the Internet. When people think of e-commerce, they often think of retail e-commerce, typified by companies like *Amazon.com*. Most of the sales growth on the Internet, however, has been in the area of **business-to-business (B-to-B)** e-commerce, rather than in retail **business-to-consumer (B-to-C)** e-commerce. Let's look at B-to-C e-commerce, then at B-to-B e-commerce.

Business-to-Consumer E-Commerce

There are many reasons why companies want to use e-commerce to sell their products to the public:

- E-commerce is a good way to reach customers. By 2004 B-to-C e-commerce sales are expected to reach $180 billion.

- Web transactions allow a company to use data-mining techniques to record and analyze information about its customers. With that information, a company can improve marketing, thus making more sales and increasing profit. For example, when a customer goes to *Amazon.com,* she is presented with product suggestions that are based on her previous purchases.

- E-commerce allows companies to be more efficient, streamlining operations and lowering costs. Streamlining operations often begins with changing the way in which a company interacts with its customers. For example, instead of employing

130

customer service agents to take telephone orders, customers can order directly from the Web, check their orders online, and ask questions via e-mail. As a result, customer service is enhanced and costs are lowered.

- E-commerce provides customers with incentives to order. For some buyers, the incentive is convenience: Ordering is quick and easy—and it can even be done from home. Other buyers are attracted to unique services. For example, one company created a toy "wish list" during a holiday season. Parents and relatives could view a child's online wish list and make their purchases directly from the list. Other companies (Lands' End, for example) offer a "digital dressing room" on their site where customers can see clothes on figures shaped like their own. Still other companies offer "Web only" bargains.

- The Web gives small companies access to huge markets that once were only available to large companies. Using Web-based commerce technology is rapidly becoming a necessity, however, as competitors adopt the same practices. For example, offering online banking was once a competitive advantage, but now most banks offer the service as a competitive necessity. In fact, customers now *expect* sellers to be on the Internet.

Customers also use the Web to find better prices and better price information. Currently, some Web search engines go out on the Internet and find the lowest price for a particular type of item. For example, *CNet.com* provides this kind of price comparison for computers and peripherals.

Business-to-Business E-Commerce

Business-to-business e-commerce is defined as buying and selling between two companies over the World Wide Web. The companies might be manufacturers, suppliers, wholesalers, or retailers. Estimates of B-to-B e-commerce sales are currently about $400 billion per year and are expecting to reach $7 *trillion* by 2004.[1]

Business-to-business e-commerce is transforming how companies sell to each other. For example, auctions allow companies to buy raw materials at the lowest possible cost. These auctions are run like *eBay.com*, but they are specific to a particular industry. So, for example, when a company's purchasing manager wants to buy sugar, she can go to an auction site— either for the food industry or a general "wholesale/retail only" auction site—and click on "food products."

Electronic Data Interchange (EDI)

B-to-B is not new. Prior to the invention of the World Wide Web, companies electronically transferred purchase orders through a system known as **Electronic Data Interchange (EDI)**. EDI is an electronic computer-to-computer transfer of standard business documents. According to one study, EDI trading volume was nearly $3 trillion in 1999.[2]

Companies can set up their own networks to communicate with their suppliers' systems, or they might subscribe to a network run by an outside EDI service provider, a **value-added network (VAN)**. When a company runs low on a specific raw material, it can order materials electronically. EDI should not be confused with a simple e-mail message that reads, "Dear Joe, please send us 5,000 widgets." Rather, the message is sent on a standardized business transaction form, following a computer protocol in North America known as the ASNI X.12.

The benefits of EDI have been enormous:

- Costs of paper, printing, and postage have almost disappeared from ordering systems.

- Errors have been brought to a minimum because orders do not have to be rekeyed into a supplier's system.

- Ordering is fast and efficient. As a result, large companies have forced their smaller suppliers to use EDI, via a VAN if necessary. The supplier pays the VAN per amount of information sent and received.

- Suppliers and buyers are "locked" into business relationships: Once a company sets up an EDI system with its supplier, it takes effort to change suppliers, and most do not, hence locking in the relationship. This is an advantage as long as both suppliers and buyers remain satisfied.

Internet-Based Procurement

Even though EDI has been useful, companies are moving from EDI to Internet-based procurement for the following reasons:

1. The Internet is a lower-cost network than private EDI networks.

2. Purchasing costs are further reduced as suppliers compete for orders on the buyer's Web site.

Thus, locking in suppliers is not the same and may not take place at all in Internet-based procurement. Suppliers that cannot compete on price may fall by the wayside, leaving the buyer with relationships with only viable suppliers. Here are three recent examples of this phenomenon:

- In 2000 Ford Motor Company and General Motors announced that they are putting together an enormous Web site called a "B-to-B automotive trading exchange." This venture is in conjunction with Commerce One and Oracle, which is providing the necessary software. Each company, along with its suppliers, can buy and sell automotive parts on this Web site. Purchasing auto parts will be fast and efficient, and auto manufacturers can maintain a low inventory. These savings could translate to more than 10% of the purchase price of a car. Other industries, in addition to the auto industry, are setting up exchanges.

- ERP companies like SAP have also entered the B-to-B e-commerce market. SAP has proposed a joint effort with Statoil, the world's second-largest supplier of crude oil, to develop a vertical marketplace for oil and gas services. SAP's *mySAP.com* supports exchanges in buying and selling raw materials with large companies like Statoil as partners. Statoil is linking with SAP and offering a Web site to find all types of equipment and services for the oil and gas industry. SAP hosts the marketplace, which will provide sources for services, equipment, and maintenance to the oil and gas industry. Statoil expects large savings by using a single marketplace. Other companies are expected to use the marketplace as well. As the long-established oil and gas industry has matured, pricing has become highly competitive. Any product that makes the industry more efficient is welcomed.

- The World Chemical Exchange allows companies to buy, sell, and trade chemicals at *chemconnect.com*. The Web site claims the following advantages:

  1. Best prices without traditional negotiations

  2. Faster contracts

  3. Access to new worldwide markets and new trading partners

  4. Instant market information

Internet Auctions and Reverse Auctions

B-to-B e-commerce also features online bidding through auctions and reverse auctions. In a standard auction, a company might put up for bid its products or even its obsolete equipment. In a reverse auction, a company might ask for bids to supply it with goods, services, or even telecommunications capacity. Here's how a reverse auction works. Let's go back to our example company, Fitter Snacker. The company needs raw materials such as oats and wheat germ, which for FS are essentially commodities. The company can go to a "bidding" Web site and set up a program to run overnight. The program uses the Internet to solicit bids to supply those raw materials. In the morning, FS's purchasing agent can choose the lowest price offered for the oats and wheat germ.

Internet-based auctions are changing the way in which commodities are purchased. A few years ago, commodities would have been purchased through a supplier, or middleman, who negotiated prices of raw materials with suppliers. Now, the Internet and its bidding programs have threatened the middleman's role and made the buying process more efficient—in a sense, the Internet has become the middleman. Pricing is dynamic and open.

In fact, dynamic pricing is not only forcing out middlemen, but it is also putting pressure on sellers. Sellers must be nimble and flexible to participate in this fast-paced market. This requires a seller's accounting and logistics operations to be in excellent shape before trying to sell in the auction market.

As a result of these market changes, companies are seeing efficiency gains through Internet buying and selling. For example, General Electric's appliance division has seen transaction costs drop to one tenth of what they were with telephone and fax orders. With 45% of the division's sales on the Internet, that relates to a whopping $4 billion in cost savings and $3 billion in sales growth in 2000.[3]

This success is not without a downside: No company (or government agency, for that matter) is totally immune to security breaches. In 1999 some large trading firms were shut down due to denial-of-service attacks—hackers were bombarding the sites with so many messages, the site couldn't handle the volume. Security is an ongoing effort. E-commerce companies use virus-scanning software, encryption, intrusion detection, and other measures to protect their sites and the privacy of customer data.[4]

E-Commerce and ERP

You might ask yourself, what does e-commerce have to do with ERP? The answer is that each technology complements the other, and each is necessary for total success. Without e-commerce, a company cannot compete with companies offering similar goods. Without ERP, a company cannot fill Web orders expeditiously. Here's why.

When a company receives an order through its Web site, the company can't merely file or print orders for later handling, like e-mail updates and order tracking. The orders must be efficiently fed into the company's marketing, manufacturing, shipping, and accounting systems—a series of steps sometimes called **back-office processing**.

An efficient back-office operation is crucial for any company's success. E-commerce often exacerbates problems and reveals weaknesses in current back-office systems. For example, *Amazon.com* invests its cash flow back into warehouses and other support that the company needs to keep its back office in order, so it can continue its success. Some Web-only businesses are frightened that companies like Wal-Mart will enter the Web market. Why? Because those companies already have well-established, integrated back-office and distribution systems.

Some companies with unintegrated information systems have built Web sites before creating an integrated back-office system. As a result, those companies often can't fill orders in a timely fashion, as was apparent with many e-commerce businesses during the 1999 holiday season. One large toy company announced less than a week before Christmas that it would *not* be able to fill all its Web orders. Surprisingly enough, all the toys were in the warehouse, but the company couldn't organize basic functions—like picking, packing, and shipping—to get toys to consumers on time. So integrating the Internet front-office operation and the ERP back-office operation is fundamental in today's business environment.

FS's Web-Ordering Problem

FS has traditionally sold its NRG bars through the sales force in either the Direct Division or the Distributor Division. With many companies moving to e-commerce, FS's customers want to order directly from a Web site.

Let's assume FS has implemented neither a Web-ordering system nor an ERP system. In the current annual-planning cycle, FS executives are looking at two IS investment options: (1) implement a Web-based ordering system, or (2) implement an ERP package, like SAP R/3.

Company executives would like to make both investments, but they're reluctant to make two large investments in one year. They consider both investments, weighing integrated operations with ERP against customer pressure for Web-site ordering. Executives succumb to customer pressure and implement a secure Web site so customers can place orders online. Initial demand is unknown. To promote sales and be competitive with other online merchants, the company promises delivery in five working days, which seems reasonable, based on FS's past performance.

After the Web site's implementation, orders come in more frequently (which is good—business is thriving), but orders now have shorter lead-times, and the company's back-office operations are strained. Here's what's happening: Web orders are saved to a text file, printed, and then electronically transferred into the company's own order-entry system, as described in Chapter 3. The order is then handled using the warehouse procedures discussed in Chapters 3 and 4.

The retailer who orders NRG bars from the Web site *thinks* he will be receiving the bars within five business days, as stated on the company's Web site. After 24 hours has elapsed, however, the customer receives an e-mail advising him that the NRG bars are out of stock and will not be available for at least one week, maybe two. The company's IS, barely adequate for personal and phone-based ordering, is simply too slow for Web ordering. With e-commerce, time is now compressed, and FS's unintegrated system cannot keep up with this pace of business.

FS executives now wish they had implemented an ERP system. An ERP system, with its integrated software modules and centralized database, could tame this out-of-control problem. With ERP implemented and connected to a Web-ordering system, orders on the Web could flow directly into the ERP system. When customers order NRG bars on the Web, they could see the first-available shipping date; thus advised, customers could make an informed choice.

Recent studies on back-office systems concluded that an attractive Web site is not enough for a "dot com" business to stay afloat. The conventional back-office systems must be in place and operating correctly for the business to flourish. Infrastructure is key for e-commerce success.

Exercise 6.1

1. Do research on the Internet or use library sources to find out how to implement an e-business. List the steps involved. Report your findings to the class.

2. Do you think customers' expectations are different when ordering on the Web versus ordering in a bricks and mortar store? Use some of your own experiences to answer this question.

3. Referring to the FS example, trace the steps needed to fill a Web order. Be sure to include how the ERP database interfaces with the Web page.

4. Research e-commerce businesses during the holiday seasons of 1999 and 2000. What was the common theme among companies that failed or did poorly? You can conduct your research on the Internet.

5. Refer to Figure 2 in Chapter 3 and review the six-step sales cycle. What elements in the icons are not present in a sales order received over the Web?

■ Using ERP Through an ASP

Many companies today outsource their operations to a third party. For example, the bank J.P. Morgan & Co. outsources all of its IS work. Next, let's look at a special kind of outsourcing.

Application Service Providers

An **application service provider (ASP)** is a company that delivers software applications to a company over a network. Usually that network is the Internet.

Using an ASP to get software is analogous to leasing a car. For some people, there are great advantages to leasing a car. First, leased cars are usually new and few of them break down, so leasing offers low maintenance. Second, leasing provides a way to have a car to drive without the obligation of the high purchase price. Third, leasing allows you to have a *new* car (or in software, the latest version). Finally, the cost of leasing is constant and predictable.

The company using the ASP does not own the software or the hardware and does not employ the people running the system. The ASP owns the hardware and the rights to ERP software, and it employs the workers. For example, suppose an ASP were to run an ERP package like SAP R/3 for FS. The ASP would own the rights to use the SAP software. The ASP would also own the server on which the SAP software runs. Because an ASP would tailor a software package to suit the company, FS would be relieved of finding ERP specialists. The ERP program would be delivered to company workers via a network, perhaps the Internet. The ASP would charge FS either a monthly or yearly fee, or it could charge a per-use fee for the system, depending on the number of users and the number of modules. Thus, an ASP offers a lower cost of entry for utilization of software that some companies cannot otherwise afford or maintain.

ASPs are gaining popularity. Experts predict that the market for ASPs (for all types of software) will exceed $22 billion in 2003.[5] Let's explore the pros and cons of this new way of delivering software and applications.

Advantages of Using an ASP

Many companies are finding it advantageous to use an ASP for ERP or other IS applications. Some of those benefits include the following:

- **Affordability**—Companies that once thought they couldn't afford an ERP system can now pay for it on a monthly basis. Gaining access to an ERP system through the Internet is even cheaper than using private leased lines. ASP services can be received through an Internet connection using a regular browser and the SAP graphical user interface.

- **Shorter implementation times**—The time for implementation is shorter for those who choose to implement ERP through an ASP. ASPs already have similar systems running and are experienced in software delivery. They have servers, telecommunications, and highly trained personnel already in place.

- **Expertise**—ASPs are experts in delivering IS applications. They do all the maintenance, including backups, training, and customizing of the system. That lets their customers eliminate IT headcount to focus on their business. ASPs can also run information systems more efficiently because they do it on a large scale. ASPs can spread fixed costs over many users, thus achieving economies of scale that might translate into low lease rates. And, in this tight IS labor market, an ASP may have a better chance of attracting and retaining a talented work force.

Disadvantages of Using an ASP

There are some potential problems with using an ASP:

- **Security**—Companies using ASPs are turning their information systems over to a third party; therefore, they must be confident that the ASP has a high level of security. How hardware will be shared is also a security concern: The ASP will have multiple users on a single piece of hardware. Each customer's data must be shielded from other customers. An ASP might have better security, however, technology than a small company can obtain.

- **Bandwidth/response time**—The telecommunications channel from the ASP to its customers must be fast enough to handle multiple users. An ASP's servers must be sufficient in terms of processing capabilities.

- **Flexibility**—An ASP should be flexible in working with its users and satisfying their requests for processing modifications.

- **No frills**—An ASP can usually provide basic systems well, but asking for unusual configurations may cause problems. An ASP might not allow for third-party add-ons either. Further, the ASP might not want to do much customization using the SAP R/3 programming language, ABAP.

- **Technical, not business**—An ASP knows the technical aspect of the software, but it will need the customer to define the business processes and make the configuration decisions.

Other Considerations

As with all forms of outsourcing, companies considering an ASP should carefully scrutinize the ASP's contract before signing. There may be hidden costs and potential problems.

Interestingly, ERP companies are excited about ASP capabilities. Why? Because ERP vendors—like SAP—can provide ASP services too, and that means extra profit and a steady income. Prior to delivering software through an ASP, companies like SAP relied on a small number of very large sales to make their yearly profit. Now income is steadier because SAP can deliver software on a monthly basis. For example, SAP might have a great third quarter of a given year because it made some very large sales, but the next quarter it might not sell another such package. By delivering software through an ASP, however, the lease payments will provide SAP with a steady monthly income. SAP's customers benefit, too: SAP says that in a small company, SAP R/3 delivered via *mySAP.com* can be up and running in a mere five days!

Exercise 6.2

Let's return to our example company, Fitter Snacker. Assume that FS has made the decision to acquire an SAP system. The newly appointed CIO must decide how to implement the decision. After talking with the largest vendor of ERP software, SAP, the CIO realizes FS has two options for implementing ERP:

1. Buy new computers, disk space, the rights to the software, and a full installation.

2. Run FS's ERP system off the Web via an ASP, which would deliver SAP R/3 for a monthly fee.

There are pros and cons for each type of implementation. In this exercise, you must recommend one course of action to FS's CIO. The first part of this exercise is to write a memo to the CIO, enumerating the pros and cons of each method of implementing ERP. Use the following table to organize your thoughts:

FIG. 6.1

Advantages of purchasing software and computers for ERP	Advantages of using an ASP to run ERP

After passing around your memo to various board members of FS, the treasurer of FS needs to justify his decision. He would like to see a spreadsheet that will crunch the two scenarios' data and come up with a financial decision. Your job is to compare the monthly cost of using the ASP with the capital expenditure of investing in an in-house ERP system. Weigh the pros and cons of each method, and then make a recommendation to the treasurer of FS. Here are the details of the decision:

Option 1: Buying Computers and Software Rights
Here is what FS must buy to set up its SAP system:

■ Database server—A UNIX server would cost $70,000.

■ Application server—FS needs a server to run the SAP application, which would cost about $40,000 since there aren't any servers available for use in the system. (SAP is platform-independent; therefore, it can be run on different types of computers. Thus, a company can often use an existing server.)

■ PCs—Some of FS's existing PCs could be loaded with the SAP graphical user interface (GUI) and access the system. Because more FS employees will be connected to SAP R/3, however, they will need 10 additional computers. Total cost for the PCs would be $15,000.

■ Rights—Multiyear rights to SAP software for all users would be $500,000. The treasurer is unsure if further outlays will be required after the fifth year, and therefore is limiting the analysis to the years 2001 through 2005.

■ Installation—SAP will help, but it also needs to hire consultants for the six-month implementation. At $3,000 per day, the whole cost would be an estimated $486,000.

■ User training—With the purchase of the rights to the SAP software, FS receives SAP training at a local SAP training center. This training is for key personnel involved in the project. FS feels it needs further training, however, for FS-specific business practices, which will be for two weeks at $2,000 per day. This includes a training consultant to run classes at FS headquarters. The cost of training would be $20,000, plus travel and lodging, which brings the total number to $23,000.

- Ongoing consulting—Once the system is up and running, FS will need to pay for consultants to help maintain the system. FS has chosen SAP as its consultant, so if FS runs into trouble with the software, FS will have the best advice. FS estimates that if it budgets for consultants to come in once a month for $3,000 a day, it should be able to have all employees' questions answered. The total yearly cost of additional consulting would be $36,000.

- Computer maintenance—FS needs to make sure all PCs and servers are running properly. To do this, FS would purchase a maintenance contract to cover all hardware. This contract would cost about $1,000 a month, or $12,000 yearly.

- Network and database administrator—FS would need a full-time network and database administrator to run the system. Salary, including benefits, for a skilled person would be $200,000 per year.

Option 2: Using ERP Through an Application Service Provider

The other option is to use an ASP. Estimated costs for this option are as follows:

- PCs—FS still estimates that they would have to purchase 10 more PCs because many users will now be accessing the computer system. Each PC would cost $1,500, or a total of $15,000.

- Maintenance on PCs—The maintenance contract on all the PCs would cost $600 per month, or $7,200 yearly.

- Software through the ASP—The monthly cost of delivering SAP to FS via the Web would cost them $33,333, or $400,000 yearly.

- Training would be provided by the ASP.

Make a recommendation to FS. Which option should FS choose—purchase the SAP software and computers outright, or use an ASP? Set up a spreadsheet that will add all the costs of each option. In each scenario, you must deal with the **net present value (NPV)** of money.

NPV is a way to figure out whether an investment is profitable, or in this case, compare outlay of funds from one method to another. NPV can be calculated over a number of years; in our case, a five-year outlay of funds for the ERP project. The syntax of NPV is =NPV (hurdle rate %, range of values) within an Excel spreadsheet. The values in the range can be positive or negative numbers. In our case, they are all outflows, but we can work with them as positive numbers. The **hurdle rate** (written in the spreadsheet as a decimal, not a percentage) is the rate of discount over the period. The hurdle rate is the minimum acceptable rate of return on a project that a company will accept.

Your spreadsheet should begin like the one shown in Figure 6.2 (with years continuing through 2005).

FIG. 6.2 Cost comparisons: buying versus renting

Purchasing Options							
Option1 - Buying computers and software outright							
Items			**2001**	**2002**	**2003**	**2004**	**2005**
Database server			70000				
Application server			40000				
10 PCs			15000				
Software - SAP			500000				
Consultant - initial (6 months)			486000				
Training (2 weeks)			23000				
Consultant - maintenance (1 day per month)				36000	36000	36000	36000
PC maintenance				12000	12000	12000	12000
Network administrator			200000	200000	200000	200000	200000
Total			1336001	250002	250003	250004	250005
NPV			$64,214.34				
Option2 - Using an ASP							
PCs			15000				
PC maintenance				7200	7200	7200	7200
ASP cost			400000	400000	400000	400000	400000
Total			415000	407200	407200	407200	407200
NPV			$20,731.42				
Hurdle rate			20				

To complete this exercise, perform the steps that follow:

- Calculate the cost of the two methods of implementing SAP R/3 for five years. Use the spreadsheet as illustrated in Figure 6.2 as your guide. Use the NPV calculations to reference the hurdle rate at the bottom of the spreadsheet. Vary the hurdle rate, following the directions your instructor provides.

- Consider using different hurdle rates for each option. Why might that be applicable for this decision?

- Write a memo, with your spreadsheet attached, to the treasurer. Answer this question: Which method should FS choose—and why? Be sure to consider both the qualitative aspects and the quantitative aspects of the choice. Also address the viability of the ASP.

Your instructor might assign the following additional exercises:

1. Use the Web or library resources to research the use of ASPs. Find cases of companies that have been successful in using an ERP system via an ASP. Describe one success story in a memo to your instructor.

2. Would you give different advice to a smaller company than you would to a medium-sized company, or to a large company, regarding the use of ASPs? Why?

■ Connectivity

Today's business environment is one of constant change. Companies with better access to information have a competitive advantage over those that do not. Telecommunications has developed and flourished because it overcomes the physical separation of workers and therefore increases efficiency. To be connected any time from any place to the back-office system can provide that competitive advantage.

Types of Connectivity: Web and Wireless

People are connected all over the globe through the World Wide Web. The Web allows connectivity from anywhere on any type of computer connection because the transmission formats are the same for all. **Internet service providers (ISPs)** are permanently connected to the Internet. They agree to forward traffic passing through the network. Thus, using the Web allows a company to communicate with customers and suppliers in one medium, the Internet, no matter what type of computer or operating system is used.

Telecommunications has evolved rapidly. For example, in the early 1990s, workers would check information or e-mail from their office's IS by dialing into a computer, using a telecommunications connection (often seen in airports and hotels as data jacks). Now many organizations are using e-mail systems like *Microsoft Outlook*, which allow the user to log on and then process e-mail from any Web browser and Internet connection.

Using a Web browser and a fixed Internet connection is not the only way to remotely access a company's IS. Many users are now choosing wireless technology for this connectivity. Europeans have been quick to adopt the wireless technology because of a wireless standard already in place. The United States is expected to follow that standard. The number of wireless devices in North America is expected to double from 2000 to 2004.

Advantages of Connectivity

Many people must, or are choosing to, work away from their offices. There are many advantages of having these employees universally connected to their company's IS:

- **Efficiency**—For the traveling representative, connectivity means faster response time to a customer's order or request.

- **Productivity**—More time is productive since access to the system can be made from virtually anywhere.

- **Data collection**—The user can easily collect data in the field and transmit it to the office, which can mean connecting to the company's back-office system.

- **Profits**—Profits may eventually rise with the increased efficiency and productivity.

For some companies, connectivity is a competitive necessity. Pressure from customers and competitors leads companies to invest in this technology. Connectivity not only connects workers to their office, it also helps customers to order online and vendors to sell online. Efficiency is increased because orders are being placed directly into the system; rekeying of data does not occur.

Companies are using connectivity to boost profits in many ways:

- Barnes and Noble provides a wireless connection to its e-commerce music site. Customers can listen to CDs and place orders without any physical connection.

- Continental Airlines lets passengers receive up-to-date flight information over wireless devices.

- J.P. Morgan lets users get price quotations on bonds as well as buy and sell them via wireless connection.

ERP vendors are also addressing the issue of connectivity. For example, SAP offers a system solution called *mySAP.com*, which allows access to the system anywhere in the world via a Web browser and an Internet connection. It also includes access to one's own ERP system via third-party wireless devices.

Wireless Connectivity Challenges

As with any new technology, there are challenges facing wireless connectivity:

- **Cost**—Outfitting all workers with wireless devices, buying or renting air waves, and setting up headquarters communications devices can be expensive.

- **Coordination**—Coordinating all the devices on a wireless network is challenging for network administrators.

- **Current lack of standardization**—Many standards exist in North America, and that causes a technical challenge. For compatibility, companies must control equipment purchases.

- **Security**—Wireless transmissions are susceptible to interception. Security measures do exist, including encryption.

FS's Web-Access Problem

To illustrate this problem, let's assume Fitter Snacker now sells internationally. Business is going very well, and the salespeople are very busy. Most of them travel out of the office at least four days a week, and they travel nationally and internationally.

To save time and money, FS has adopted the SAP Travel Planning Application, which helps to manage travel services and expenses for the company. The system can record all travel expenses and store them in a centralized database, thereby allowing faster approval and reimbursement. One great convenience of this system is the real-time access the corporate traveler has to reservations, weather reports, entertainment information, and mapping services. This access can be via a personal digital assistant (PDA), such as a Palm Pilot. So when an FS sales representative is delayed in a traffic-jammed taxi, he can use his PDA to change a reservation at a hotel or on an airplane. The Amadeus Company has teamed up with SAP to bring the SAP Travel Planning Application to the R/3 version 4.5 release. Some advantages of this system are as follows:

- Real-time access to hotel, airline, and car rental reservations

- Recording, and therefore tracking, of all travel expenses in a data warehouse

- Better account management of travel expenses

- Lower-cost contracts with travel service providers. Now that the company has a complete understanding of all travel expense data, volume discounts can be negotiated.

Exercise 6.3

1. Assume you are the top salesperson at FS and spend about 75% of your time traveling. Write a memo to the CEO of FS, justifying your use of the Amadeus travel system. Be sure to list the benefits and drawbacks, if any, to the system. Cite possible examples of use to justify this system.

2. Conduct Internet research or use library resources to find how companies, in particular those using ERP, have used connectivity to improve efficiency. Try to find quantitative data about how profits have increased by using this new technology. Summarize your findings in a brief memo.

3. For many years, economists were reluctant to say that information technology had contributed significantly to increases in business productivity. A 2000 report from the Federal Reserve states that IT has contributed to the economy substantially. Now many well-known economists say that IT has been the main factor in the United States' sustained economic boom. Whatever the studies say, productivity is hard to measure, and many companies don't do it because it is just too difficult. Discuss whether wireless technology makes ERP more productive. Research the Internet for supporting articles—for example, surveys done by *Information Week* magazine.

4. Hold a classroom debate on cost-benefit analysis. You can use FS as an example. You will also need to research your topic to prepare your side's arguments. Your instructor will divide your class into teams to debate the following topic:

   Resolved: Companies should not do a cost-benefit analysis when deciding whether to invest in and implement a new connectivity system for their existing ERP system.

Background: Often, a reliable cost-benefit analysis is not possible with ERP systems. Some companies have abandoned doing cost-benefit analyses because of a number of factors:

- ERP software is so expensive, the payback is beyond the range of years normally used for cost-benefit analyses.

- Benefits are intangible. How do you put a price on integrated information versus unintegrated information?

- Many returns on investment are negative. Normally, only projects with positive ROIs are funded.

Another Look at the Future of ERP

The following is the text from an interview with Claudio Spiguel, Vice President of Commercial Information Management for AstraZeneca Pharmaceuticals in Wilmington, Delaware.

What are the perceived benefits of ERP for a company like yours?

Business in the past 10 years has gotten more complex and now has many dimensions. ERP can help our company in two major areas:

1. Globalization: Different countries have different ways of doing business. ERP can replace those global differences.

2. Regulatory environment: In our business, there are strict regulatory rules, and they are different in different markets. ERP can help manage company-wide compliance with all those complex regulations.

ERP is basically using systems to coordinate the flow of information to maximize profits. You need to strike a balance, however, between total integration and nonintegration.

There can be problems with total integration. The complexity of business is so high that implementing a total ERP system for a large multinational company is very difficult. One needs to approach ERP with reality in mind. For example, one of our factories in Brazil had an ERP system in place. It worked well and the factory was profitable. It was not a good business decision for them to change their non-SAP system to our headquarter's SAP system, so an interface was written to bridge the two systems.

Why is ERP an easy sell to top management?

There is a one-word answer to that question: control. All upper-level managers want to control their organization. They often do not have access to proper information at the proper time. Any system that can give them accurate, timely information is fantastic. In fact, years ago, we implemented the old SAP system, R/2, just for that purpose. The workers in our factory were incredulous because we went from a Windows-based graphical user interface to a text-based interface with R/2. To them, it seemed like a step back in technology, but that new system could give upper-level management the information they needed to control the company's operations.

Are there any unusual concerns for ERP in the pharmaceutical industry?

In the pharmaceutical industry, certain forces lead to fragmented systems, not integrated systems. For example, an individual customer's genetic code is one type of information that would not be widely available on an integrated system due to privacy concerns. Also, in this regulated business, ERP systems have different flows than that seen in the traditional paper-based system that the FDA [Food and Drug Administration] still uses. The FDA is moving into the electronic age, but government regulators don't move at the same pace as industry.

How is e-commerce changing the face of ERP in your industry?

E-commerce is forcing business into an environment that is faster than ever and more uncertain than ever. One interesting part of the e-commerce puzzle in our industry is the area of electronic medical records. The concept of integrated information for medical records has appeal for all healthcare practitioners, insurance companies, and pharmaceutical companies. Freely available electronic patient records, however, are the last thing the public wants. So although the concept of a fully integrated system for medical records is appealing to the industry, it is not now possible due to privacy concerns.

For example, assume that all clinical trial data were in an ERP system, accessible by many types of users such as pharmaceutical companies, doctors, and researchers. Cost savings and research advances would result. Many users would like this, but the patients in the clinical trials would be concerned their privacy was being invaded.

What is the future of ERP?

It is true that the back-office systems need to be in place for e-commerce to take place successfully. E-commerce is impossible without the infrastructure ERP provides. ERP faces risks in the new electronic environment, however, because of the Web's reach, speed, and lack of time to retract errors. Time is now compressed and mistakes that happen can be big. So there is uncertainty in the future of ERP and for all businesses. Business people will have to work hard to respond to the speed at which ERP and the Web can deliver information.

Questions

1. The interview raises issues about ERP leading to a lack of personal privacy. Do you think that the "greater good"—society's cost savings in medical insurance and research—should outweigh the individual's right to privacy? Debate your opinion with classmates.

2. Dr. Spiguel states that government regulations don't keep pace with industry's technological advances. Why do you think that is the case? What might be the outcomes of such mismatches in IS sophistication?

3. Another problem with medical records and research on the Internet is that once incorrect information is released over the Web, correcting the information is very difficult. In fact, there have been many cases in which medical research findings have been misinterpreted and then propagated over the Web. Incorrect information in other areas has also been spread over the Web. Use the Internet or library sources to research incidents of misinformation being spread over the Web. Were there any consequences to the misinformation? Share your findings with classmates.

4. Dr. Spiguel is an IT vice president in a large for-profit corporation. Pat Davis (Chapter 5) is an IT project director for a midsized university. You have read the interviews with each person. The aspects of ERP with which they are concerned are very different, as you can see from the interviews. What are the major issues with which Spiguel and Davis are concerned? Point out differences and similarities (if any).

5. How do you think a person's role affects his or her view of ERP? Justify your views by referring to what Spiguel and Davis talk about in their interviews. To answer this question, you might consult MIS or systems analysis and design texts for CIO and systems analyst job definitions.

Another Look at E-Commerce

Chocolate giant Nestlé will spend the next three years rolling out SAP R/3 to its largest facilities, with other smaller facilities to follow. The European icon has chosen *mySAP.com* to deploy to its 230,000 employees in 500 sites in 80 countries. Initial cost of the venture is $200 million. There is also up to $80 million being allocated for consulting, maintenance, and upgrades.

In order to compete in the e-commerce market, Nestlé realized that they must get their back-office infrastructure in order. Currently, they have a "mishmash" of systems:[6]

- 900 IBM AS/400 midsize computers

- 15 mainframes

- A variety of operating systems, including UNIX and various versions of Windows

- Some sites already running SAP R/3

The systems are not integrated with headquarters. Data are transmitted to headquarters via an EDI network or slow dial-up lines. Nestlé will install SAP R/3 and upgrade its computing and telecommunications hardware. Employees worldwide will interface with the ERP system via *mySAP.com*.

One feature of the *mySAP.com* system is that each employee will have a customized interface to the system for his or her specific job. The system is accessed via a browser. SAP has specified the homepages for each job function, which requires the employee using it to follow its business practices. Nestlé claims that theft has already been reduced with the new system, which rigorously ensures that each employee does everything correctly, in the correct order. Previously, some employees had submitted fraudulent invoices. Security is also increased since each employee only has access to the work required for his or her specific job.

There will be five data centers internationally. For the first time, management at headquarters will be able to see company-wide aggregate data. For example, sales information will be used to reduce overstocking of shelves and subsequent spoilage of products.

Nestlé expects that employees will be unhappy about the transition to the new ERP system. Nestlé is providing extensive training, including one-on-one training, classroom training, and computer-based training.

Questions

1. Why might some Nestlé employees be unhappy about the new SAP system?

2. Besides providing software training, how else can the company ease the transition to the new system?

3. What are the benefits of each employee having his own "portal" to the system, tailored for his specific job?

4. List all the benefits you see of implementation. Draw ideas from the previous chapters as well as from Chapter 6.

▉ SUMMARY

1. E-commerce is transforming the way companies do business. Business-to-consumer e-commerce can streamline a company's ordering operations and record information about customers, making the company more competitive.

2. Business-to-business e-commerce is changing the way companies buy and sell goods. New forms of procurement such as auctions, reverse auctions, and trading exchanges—all with dynamic pricing—are replacing the middleman.

3. ERP is essential for all forms of e-commerce to provide speed and consistency to transaction processing and other back-office operations.

4. Application service providers are allowing companies to acquire ERP without a large initial investment. There are risks associated with using an ASP, however, and the decision to buy or lease must be weighed carefully.

5. Wireless connectivity is rapidly gaining popularity globally, and ERP systems are becoming connected to those wireless devices. Access to better information can provide a competitive edge.

■ End Notes

1. "The Coming B2B Explosion," *The Marketeer* (online) (February 8, 2000).
2. " The Coming B2B Explosion."
3. Pamela L. Moore, "GE's Cyber Payoff," *Business Week* (online) (April 13, 2000).
4. Kathy Williams, "How Secure Is E-Commerce?" *Strategic Finance* (March 2000): 23.
5. Audrey Apfel, "A New Lease on Software," *Executive Edge* (June/July 2000): 38-41.
6. Steve Konicki, "Nestlé Taps SAP for E-Business," *Information Week* (June 26, 2000): 185.

GLOSSARY

Accelerated SAP (ASAP): a set of tools designed to speed the SAP R/3 implementation process.

activity-based costing: an advanced form of inventory cost accounting in which overhead costs are assigned to products, based on the manufacturing activities that gave rise to the costs.

Advanced Business Application Programming (ABAP): SAP R/3's internal programming language.

application service provider (ASP): a business that delivers software applications to companies over a network. Sometimes that network is the Internet.

assemble-to-order: manufactured products that are produced using a combination of make-to-stock and make-to-order processes. This method employs assembling component parts as customers' orders are received.

Asset Management (AM): a module in SAP R/3 that helps a company to manage fixed-asset purchases (plant and machinery) and related depreciation.

back-office processing: the processing of sales orders through a company's marketing, manufacturing, shipping, and accounting systems. A company selling on the Web must have an integrated back office to keep up with orders.

balance sheet: this summary of a company's account balances includes cash held, amounts owed to the company by customers, the cost of inventory on hand to be sold, long-term assets such as buildings, amounts owed to vendors, amounts owed to creditors, and amounts that the owners have invested in the company.

best of breed: an ERP design approach in which "the best" modules from different ERP vendors are integrated into one ERP system.

bill of material (BOM): the "recipe" listing the materials (including quantities) needed to make a product.

business functions: business activities within a functional area of operation.

business process: a collection of activities that takes one or more kinds of input and creates an output that is of value to a customer. Creating the output might involve activities from different functional areas.

business-to-business (B-to-B): communication and sales between manufacturers, wholesalers, retailers, and suppliers. This communication can occur using both EDI and the Internet.

business-to-consumer (B-to-C): communication and sales between businesses and the buying public. Popularly, but incorrectly, thought of as the most common form of e-commerce.

call-center automation: a customer relationship management (CRM) activity that occurs when customers call a company for assistance with its products.

cash-to-cash cycle time: the time from paying suppliers for raw materials to collecting cash from the customer (used in supply-chain management metrics).

client-server architecture: data stored in a central computer (a **server**) are downloaded to a local PC (a **client** of the server) where data are processed. Historically, client-server architecture replaced many companies' mainframe-based architecture.

closing the books: the process of summarizing the balances in a company's financial accounts. The purpose of this process is to create financial statements for a particular period of time, such as a year, a quarter, or a month.

Controlling (CO): a module in SAP R/3 that is used for internal management purposes. The software assigns a company's manufacturing costs to products and to cost centers, facilitating cost analysis.

cost variances: the difference between actual costs and standard costs.

currency translation: converting financial-statement account balances expressed in one currency into balances expressed in another currency.

Customer Master Data: central database tables in SAP R/3 that store permanent data about each customer. Master data are used by many R/3 modules.

Customer Relationship Management (CRM): a variety of different software tools that use data from a company's ERP system to enhance the company's relationships with its customers. CRM software allows these activities: segmenting customers, one-to-one marketing, sales-force automation (SFA), sales-campaign management, marketing encyclopedias, and call-center automation.

data mining: the statistical and logical analysis of large sets of transaction data, looking for patterns that can aid decision-making and improve customer sales and customer service. Data mining is often done with data in a data warehouse.

data warehouse: a database, separate from a company's operational database, that contains subsets of data from the company's ERP system. Users analyze and manipulate data in the warehouse. Thus, they do not interfere with the workings of the database that is used to record the company's transactions.

delivery: in SAP R/3, the word "delivery" means releasing the documents that the warehouse will use to pick, pack, and ship orders.

demand management: the third step in the production planning process. At this step, the production plan is broken down into finer time units, such as weekly or even daily production figures.

detailed scheduling: one of the latter steps of the production planning process. Detailed scheduling produces plans that can be used to run shop floor operations.

direct costs: in cost accounting, materials and labor are often called direct costs of production because constituent amounts of each in a finished product can usually be identified and estimated.

document flow: the linked set of document numbers related to an order; an "audit trail."

e-commerce: the buying and selling of goods and services over the Internet.

electronic data interchange (EDI): a computer-to-computer transfer of standard business documents that allows companies to handle the purchasing process electronically, avoiding the cost and delays resulting from paper-based systems.

Enterprise Resource Planning (ERP): ERP systems help to manage business processes such as marketing, production, purchasing, and accounting in an integrated way. ERP does this by recording all transactions in a common database that is used by information systems throughout the company and by providing shared management-reporting tools.

Financial Accounting (FI): a module in SAP R/3 that records transactions in the general ledger accounts and generates financial statements for external reporting purposes.

functional areas of operation: a categorization of business activities, including marketing, sales, production, and accounting.

Human Resources (HR): a module in SAP R/3 that facilitates employee recruiting, hiring, training, and payroll and benefits processing.

hurdle rate: a company's minimum acceptable rate of return on an investment project.

income statement: a financial statement that shows a company's profit or loss for a period of time.

indirect cost: in accounting, costs that cannot be directly associated with individual products. Indirect costs include manufacturing overhead costs such as heating, lighting, insurance, and production management's salaries.

Industry Solutions (IS): a preconfigured version of SAP R/3 that includes typical configuration options for an industry— for example, automotive, chemical, and banking. Industry Solutions simplifies the configuration process.

initial fill rate: the percentage of the order that the supplier provided in the first shipment to the manufacturer or retailer (used in supply-chain management metrics).

initial-order lead-time: the time needed for the supplier to fill an order (used in supply-chain management metrics).

integrated information systems: systems that allow sharing common data throughout an organization. ERP systems are integrated systems because all operational data are located in a central database, where they can be accessed by users throughout an organization.

Internet service provider (ISP): a company that provides Internet access to subscribers.

lead-time: the sum of the time it takes a manufacturer to get raw materials or subassemblies from a supplier— sum of the time required for a supplier to receive and process the order, take the material out of stock, package it, load it on a truck, and ship it to the manufacturer.

lot sizing: the process for determining purchase and production order quantities.

make-to-order: manufactured products that are produced to fill specific customers' orders.

make-to-stock: manufactured products that are made for inventory (the "stock") in anticipation of sales orders.

Manufacturing Resource Planning II (MRP II): an evolution of MRP that integrates MRP into other company information systems, for example, purchasing and accounting. MRP II is a close precursor to ERP. *See* "Materials Requirements Planning."

Marketing Encyclopedias: CRM software that serves as a database of promotional literature about products. The information can be routed to sales representatives or customers as needed.

master production schedule (MPS): the production plan for finished goods.

Materials Management (MM): a module in SAP R/3 that manages the acquisition of raw materials from suppliers (purchasing) and the subsequent handling and storage of raw materials, work in process, and finished goods.

Material Master Data: central database tables in SAP R/3 that store relatively permanent data about materials. These data are used by SD, MM, and other SAP R/3 modules.

Materials Requirements Planning (MRP): a production-scheduling methodology that determines the timing and quantity of production and purchase-order releases to meet a master production schedule. This process uses the bill of material, lot-size data, and material lead-times.

metrics: measurements of performance; discussed in this book in relation to the effects of supply-chain management efforts.

Moore's Law: a theory named after Gordon Moore, who in 1965 noticed that a new generation of computer chips was being released every 18 to 24 months, and the number of transistors on a computer chip (and thus its performance) doubled with each new generation. He postulated that the trend would continue, and it has.

MRP Record: the standard way of showing the Manufacturing Requirements Planning process on paper.

....

net present value (NPV): a calculation of the current value of an investment's cash flows that considers the time value of money.

....

on-time performance: a measure of how often a supplier meets agreed-upon delivery dates (used in supply-chain management metrics).

One-to-One Marketing: CRM software that first categorizes a customer according to certain criteria and then offers tailored products, promotions, and pricing accordingly. For example, customers might be offered products related to their recent purchases.

open-architecture: software that allows integration with third-party software. SAP R/3 is an example of open-architecture software. The term can also be applied to hardware products.

organizational structures: the method used in SAP R/3 to define the relationships between organizational groups such as companies, plants, storage locations, sales divisions, and distribution channels.

overhead: a company's cost of operations, such as the costs for factory utilities, general factory labor, factory management, storage, insurance, and other manufacturing-related activities. Overhead is often called an indirect cost of production.

Plant Maintenance (PM): a module in SAP R/3 that allows planning for preventive maintenance of plant machinery and managing maintenance resources, so equipment breakdowns are minimized.

Production Planning (PP): a module in SAP R/3 that maintains production information; production is planned and scheduled, and actual production activities are recorded.

Project System (PS): a module in SAP R/3 that allows planning for and control over new R&D, construction, and marketing projects. This module allows for costs to be collected against a project budget, and it can be used to manage the implementation of R/3 itself.

Quality Management (QM): a module in SAP R/3 that helps to plan and record quality-control activities, such as product inspections and material certifications.

R/2 system: the precursor of the R/3 system. Released in 1978, this mainframe-based program integrated many, but not all, of a company's operations.

R/3 system: the first integrated information system released by German software vendor SAP in 1992. This ERP system contains the following main modules, which can be implemented as a group or selectively: Sales and Distribution (SD), Materials Management (MM), Production Planning (PP), Quality Management (QM), Plant Maintenance (PM), Human Resources (HR), Financial Accounting (FI), Controlling (CO), Asset Management (AM), Project System (PS), Workflow (WF), and Industry Solutions (IS).

return on investment (ROI): a ratio calculated by dividing the value of the project's benefits by the value of the project's cost.

Sales Campaign Management: CRM software that lets a company organize a marketing campaign and compile its results.

Sales and Distribution (SD): a module in SAP R/3 that records sales orders and scheduled deliveries.

Sales-Force Automation (SFA): CRM software that logs customer contacts in a company's database. Customers that contact the company can be automatically routed to the appropriate sales representative. SFA software can also be used to forecast a customer's needs, based on the customer's history and transactions. Sometimes this software is called "lead management software" because a transaction can be tracked from the initial lead to post-sale follow-up.

sales forecast: a company's estimate of future product demand, which is the amount of a product that customers will want to buy.

sales forecasting: the process of predicting the future demand for a company's products. Usually this is the first step in production planning.

sales and operations planning: the second step in the production planning process. At this step, data are evaluated to determine whether production facilities can produce enough goods to meet the consumer demand projected by the sales forecast.

SAP: *Systemanalyse und Programmentwicklung* (Systems Analysis and Program Development—pronounced S-A-P), an ERP software vendor headquartered in Mannheim, Germany.

scalability: information systems are deemed "scalable" if their capacity can be extended by adding servers to the network, rather than replacing the entire system. Scalability is a characteristic of client-server networks, but usually not of mainframe-based systems.

Segmenting Customers: CRM software that lets a company stratify customers by volume, profitability, or other measures. Scarce marketing resources can then be directed to the most desirable customers.

standard cost: the expected cost of manufacturing a product during a particular period. Standard costs for a product are established by (1) studying historical cost patterns in a company and (2) taking into account the effects of current manufacturing changes.

Stock/Requirements List: In SAP R/3, a report that shows current stock, required materials, material expected to be received, and availability of materials.

supply-chain management (SCM): sharing long-range production schedules between a manufacturer and its suppliers, so raw materials can be ordered and delivered in a timely manner, thus avoiding stock-outs or excess inventory.

total supply-chain management costs: include the cost of buying and handling inventory, processing orders, and information systems support (used in supply-chain management metrics).

value-added network (VAN): an EDI service provider. Companies acquire EDI service by subscribing to a VAN's EDI network.

Workflow (WF): a module in SAP R/3 that can be used to automate any of the activities in R/3. It can perform task-flow analysis and then prompt employees (via e-mail) if they need to take action.

◼ FOR FURTHER REFERENCE

General Information Technology
Peter W. Keen, *Every Manager's Guide to Information Technology*, second edition. (Boston, MA: Harvard Business School Press, 1995).

This book is organized as a glossary of key terms. Keen is a knowledgeable and graceful writer. He's very strong on the history of computing and the need for integrated systems. The book is a little dated, but it is still an excellent general reference. (Keen does have a more recent book by the same publisher about using the Internet.)

E-Commerce
Gary P. Schneider and James T. Perry, *Electronic Commerce* (Boston, MA: Course Technology, 2000).

This is a comprehensive and readable review of how the Internet works and how e-commerce is done.

Periodicals
CIO, IDG Corporation, Boston, MA

This magazine provides very useful information about developments and issues in the information systems field.

InformationWeek, CMP, Manhasset, NY

This periodical is a good source of current information on the IT field. It frequently contains good surveys of companies' use of types of IT.

Intelligent Enterprise, CMP Media Inc., San Mateo, CA

This trade magazine surveys the e-commerce and ERP fields, including their intersection. It is published 18 times a year. The articles are well-written and topical, with a minimum of hype for products.

The New York Times and *The Wall Street Journal*

Each of these daily newspapers covers technology and its business implications. The online edition of the Times *is free (www.nytimes.com). The online edition of the* WSJ *is not free, but the cost is much discounted from the paper-edition subscription price.*

Online Resources
www.cio.com

The online version of the periodical previously mentioned.

www.erpworld.com

This is a content site about current events in ERP. It includes "success story" pages summarizing many companies' positive experiences with ERP. The success story pages are organized by industry.

www.howstuffworks.com

    This content site describes how a myriad of technology works, from digital cameras to e-commerce to boomerangs. It is entertaining and informative.

www.informationweek.com

    The online version of the periodical previously mentioned.

The Archives

Russel L. Ackoff, "Management Misinformation Systems," *Management Science* (December 1967): B147–B156.

    This article debunks myths about what an MIS system can do for a company's management. Ackoff points the way toward information systems that provide decision support and actual control over operations. Ackoff's advice is timeless.

Thomas H. Davenport, "Putting the Enterprise into the Enterprise System," *Harvard Business Review* (July/August 1998): 121–131.

    This is an excellent summary article about what an ERP system is, the benefits of ERP, and the need to manage ERP implementation. It is clearly written and excellent for academic use.

Claudia H. Deutsch, "Software That Can Make a Grown Company Cry," *The New York Times* (November 8, 1998): 12–13.

    The article contrasts failed implementations with successful implementations and offers advice on how to have the latter in your company. This article evidently is the source of the phrase, "Getting an ERP system is like getting a corporate root canal."

David Kirkpatrick, "The E-Ware War: Competition Comes to Enterprise Software," *Fortune* (December 7, 1998): 102–112.

    This article surveys the fortunes and capabilities of companies in the enterprise software business, broadly defined (ERP, B-to-B, SCM, CRM). Much of the information is dated, but the taxonomy is still useful, and the article would be a good place to start researching such a summary today.

Jeffrey G. Miller and Linda G. Sprague, "Behind the Growth in Materials Requirements Planning," *Harvard Business Review* (September/October 1975): 83–91.

    A still-useful review of how MRP works and why a business's management would want to use MRP software. It is also useful for understanding the genesis of today's ERP systems.

INDEX